Core Geography

The Developing World

Fred Martin and Aubrey Whittle

Hutchinson

London Melbourne Sydney Auckland Johannesburg

Hutchinson & Co. (Publishers) Ltd

An imprint of the Hutchinson Publishing Group

17-21 Conway Street, London W1P 6JD

Hutchinson Publishing Group (Australia) Pty Ltd
16-22 Church Street, Hawthorn, Melbourne,
Victoria 3122

Hutchinson Group (NZ) Ltd
32-34 View Road, PO Box 40-086, Glenfield, Auckland 10

Hutchinson Group (SA) (Pty) Ltd
PO Box 337, Bergvlei 2012, South Africa

First published 1985
© Fred Martin and Aubrey Whittle 1985

Illustrations © Hutchinson & Co. (Publishers) Ltd 1985

*Design and illustration by The Pen and Ink Book Company
Limited*
*Set in Aster and Helvetica by The Pen and Ink Book Company
Limited*

Printed and bound in Great Britain by
Anchor Brendon Ltd, Tiptree, Essex

British Library Cataloguing in Publication Data

Martin, F
  The developing world. ___ (Core Geography)
  1. Developing countries ___ Description and travel
  I. Title   II. Whittle, A.   III. Series
  910'.091724   D883

ISBN 0 09 156621 5

Acknowledgements

The authors would like to thank the following people for
their help with the preparation of this book: Ambio;
Agency for International Development; Balfour Beatty
Construction Ltd; CAFOD; Centre for US-Mexican
Studies; Centre for World Development Education; Club
du Sahel; Commission of the European Communites;
Commonwealth Development Corporation; Ecumenical
Coalition on Third World Tourism; Embassy of the Arab
Republic of Egypt; Geofile, Mary Glasgow Publications
Ltd; *Geographical Magazine*; Rev. Gordon Holmes;
International Union for the Conservation of Nature and
Natural Resources; MAYC; J. McDowell; Migrant Legal
Action Program Inc; Minority Rights Group; Namibia
Support Committee; *New Internationalist*; Overseas
Administration Department; PIACT; Population Concern;
Population Reference Bureau; RTZ Services Ltd; South
Publications Ltd; *Sunday Times*; Tarmac Overseas Ltd;
Traidcraft; UNEP; UNFPA; United States Department of
Justice; War on Want; M.W. West; World Studies Centre.

Acknowledgements are due to the following, many of
whom gave other help, for permission to reproduce
photographs: Action Aid, pp.37, 76, 79, 98, 99; Alcan,
p.113; Camerapix Hutchison Library Ltd, p.121; Camera
Press, pp.31, 36, 56; J. Allan Cash, p.105; Christian Aid,
p.41; Cuban Embassy, p.45; Culpin Planning, p.89;
Egyptian Information Service, Cairo, pp.86, 88, 90; *El
Paso Herald Post*, p.82; FAO, pp.10, 12, 16, 17, 19, 22, 24,
25, 36, 37, 42, 43, 49, 50, 54, 66, 78, 79, 92; *Financial
Times*, pp.55, 70; Tony German, p.14; Government of
China, p.75; Government of India, p.73; ILO, pp.66, 81;
IPPF, p.85; ITDG, p.48; Oxfam, pp.42, 48; A.R. Robinson,
p.69; Rössing Uranium Ltd, p.102; P.H. Starkey, pp.52,
53; Tunisian National Tourist Office, pp.116, 117; Turkish
Embassy, p.22; UN, pp.19, 80; UNICEF, pp.40, 50, 51;
Unilever plc, pp.47, 60, 61; Voest-Alpine, p.65; WFP,
pp.19, 22, 24, 38, 49, 87, 90; WHO, pp.15, 32, 33, 34, 38, 72;
WHO/UNICEF, p.35; D.E. Walling, p.70; *West Africa*, p.83.

Contents

Other titles in the Core Geography series

Leisure

1 Leisure in the landscape
Introduction; Time and opportunity; Leisure and geography; The leisure industry; Transport and leisure; Land for leisure; Workback

2 Leisure in cities
Introduction; Patterns in leisure provision; The inner city; Space in the suburbs; Improving the leisure environment; Workback

3 Into the countryside
Introduction; Travel to the countryside; Patterns of use; Cannock Chase Park; Conflicts and benefits; Second homes; The nation's honeypots; Snowdon: case study; Workback

4 Problems and plans
Introduction; Landscape facts and figures; Long-distance footpaths; Parks in the country; National Parks throughout the world; Leisure planning in the Ruhr; Workback

5 Holiday environments
Introduction; The British seaside resort; The heritage town; Tourism in the desert; The Mediterranean sunspot; Walt Disney World; Workback

6 Tourism: problems and potential
Introduction; Tourist potential in the Auvergne; Tourism and the Lapps of Northern Finland; Tourism in the Third World; Workback

Work

1 The world of work
Introduction; A world view; Work and land use; The spectre of unemployment; On the move; Workback

2 Work on the land
Introduction; The farming environment; Highlands and Islands; Rural change; The development trap; Workback

3 People and industry
Introduction; Choosing the right place; Images of places; Helping the weak; Work and its impact; Keeping in touch; Jobs and planning in Glammyd; Workback

4 Change
Introduction; Decline and death of an industry; Growing and prospering; New work from new resources; Bridging the gap; Change in Oman; Workback

5 Work and environment
Introduction; Scarred landscapes; Problems and fears; Getting it right; Wasteland Earth; Alaskan oil case study; Workback

Cities

1 Where people live
Introduction; A place to live; Move to the cities; How settlements grow and decay; World cities; Home in the city; Workback

2 City systems
Introduction; Goods and services; City arteries; Rest and play; Cleaning the city; The city day by day; Workback

3 The unequal city
Introduction; Living in Calgary; New peoples; The housing sieve; Taking action; Problem estates; Different worlds; Workback

4 The changing city
Introduction; Changing the land-use; Conserving the past; The spreading city; Building new towns; Planning for change; Workback

5 People in the countryside
Introduction; People in villages; The transport problem; The market town; Rural poverty; From countryside to city; Workback

Unit 1 The poor countries

1.1 Introduction

We live in a world of information. Live pictures from all over the world give the daily news. Books, newspapers and computers pass on the growing amount of knowledge. We can only read or listen to a tiny amount of all this information.

Telling the truth

The kind of information we get depends on who provides it (Figure 1). Newspapers and the television news stress what is new and exciting. Travel brochures tell us what is good about a country. Adventure books and films aim to entertain. They do not try to be truthful.

Fact and explanation

This textbook is about some of the world's poorer countries. Although the facts are correct, it is impossible to give a complete picture of these countries. Information, photographs and diagrams all have to be selected. Even the facts can be explained in many different ways.

This is why you must think carefully about what you read. You should read and listen to as many different sources of information as you can.

Figure 1 *Sources of information*

Exercises

(Heading: *World of information*)

1 a Write down 5 things you know about any poor country such as India.

   b Where did you get this information from?

   c Do you think you have an accurate picture of life in the country you have described? Explain your answer.

2 a Pair up which kind of information you would most probably get from each of these sources:

Information	Source
facts and figures about the people	a newspaper
important political events or disasters	a geography book
new buildings, successes and progress	a country's embassy
the best places to go	a film
the problems of a country	a travel agent
people's clothes, foods and houses	an encyclopedia

   b Write a paragraph to show why different sources can give different types of information.

1.2 Measures of wealth

There is nothing fair or equal in how wealth is shared among the world's people. There is a wide gap between rich and poor. The gap becomes wider every year.

Wealth as money

How much money people earn is one way to measure wealth. Figure 1 shows average **income per head** for people in some countries. Income per head means the amount for each person. People with a good regular income are able to buy the food, goods and services they need.

Millions survive by **subsistence**, growing or making most of what they need. With little or no money, they often have to go without.

Average income per head in US dollars (1980)			
India	226	Peru	959
Indonesia	418	Colombia	1115
Zambia	527	Jordan	1307
Sri Lanka	630	UK	8222
Phillippines	655	Sweden	13 146

Figure 1   *Average income per head for selected countries*

Average income

Average income is a good **index** of the **standard of living** in a country. This means how well people are fed, their health, and what kind of conditions they live and work in.

But the figures can be misleading (Figure 2). How wealth is shared is often more important than knowing the average figures.

What you pay for

Comparing income figures for different countries is not easy. It all depends on what has to be paid for and what things cost (Figure 3). These things depend on how the government wants to run the country.

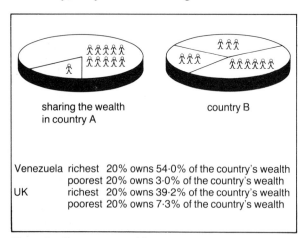

Venezuela richest 20% owns 54·0% of the country's wealth
poorest 20% owns 3·0% of the country's wealth
UK richest 20% owns 39·2% of the country's wealth
poorest 20% owns 7·3% of the country's wealth

Figure 2   *Sharing out wealth*

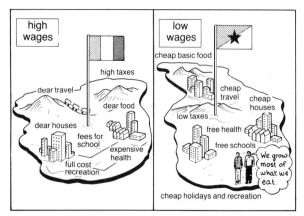

Figure 3   *What people pay for*

Exercises

(Heading: *Measures of wealth*)

1 a Explain the term: income per head.
   b What is an Index?
   c What does standard of living mean?

2 Write a paragraph to describe differences in income for the countries listed in Figure 1. Mention these things in your answer:
   the richest and the poorest
   the difference in income levels
   how people earn a living in different countries.

3 a Study Figure 2 to explain why the average income figure for a country can be misleading.
   b Compare living in the two countries shown in Figure 3. Give reasons why you would prefer to live in one country or the other.

Wealth in a country

Wealth in a country comes from activities such as farming and mining, industry and services (Figure 4). These things make up the **economy** of a country. Different types of work are called **sectors** of an economy. The terms **primary**, **secondary** and **tertiary** are used to describe the different sectors.

Wealth also comes from selling goods abroad (**exports**). Goods bought from other countries (**imports**) means that money leaves the country.

The GNP

Add up the value of everything made in factories in a year. Then add on the value of all farm produce and money earned by the services. This gives a figure called the **gross national product** (GNP) (Figure 5).

Total GNP shows how rich a country is. GNP per head shows how much each person is producing. In poor countries people work hard, but with little power and machinery they cannot grow or make as much as people in rich countries.

People at work

Figure 6 shows how the **structure of employment** is different in rich and poor countries. In poor countries, farming is still the main type of work. A small percentage of people work in factories and a large percentage work in the services. About one third do not earn a wage for the work they do.

	poor countries		rich countries	
40%	𝘏𝘏𝘏𝘏	farming	10%	𝘏
20%	𝘏𝘏	industry	40%	𝘏𝘏𝘏𝘏
40%	𝘏𝘏𝘏𝘏	services	50%	𝘏𝘏𝘏𝘏𝘏

Figure 6 *Employment structures*

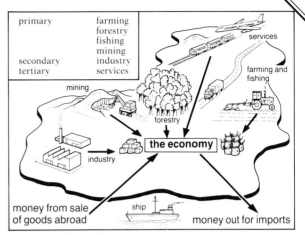

Figure 4 *The wealth of a country*

country	population in millions	total GNP in US $ 000 millions	GNP per head in US $
Japan	115	884	7700
UK	56	353	6340
Canada	24	228	9650
India	658	125	190
Chile	11	18	1690
Indonesia	139	52	380

Figure 5 *GNP for selected countries*

Exercises

(Heading: *Measuring the economy*)

4   Copy the diagram which shows how the economy of a country is made up (Figure 4).

5   a   Write a paragraph to explain how the GNP figure can help show how wealthy a country is.
    b   Why is it that people in poor countries work hard, but do not produce as much as people in rich countries?

(Title: *People at work*)

6   a   Draw a diagram to illustrate the information in Figure 6. Choose from the following types of diagram: pie charts, divided bars, pictograms (drawings).
    b   Complete these sentences.
        In poor countries, most people work at _____. Few people work in _____. About half the people work in the _____ sector.

Cars as wealth

Another way to measure wealth in a country is to count the things people own such as cars (Figure 7).

Owning a car means that a person earns well above the world average. High car ownership in a country means that there are good roads, garages and oil refineries. There will also be skilled mechanics to repair and even make cars.

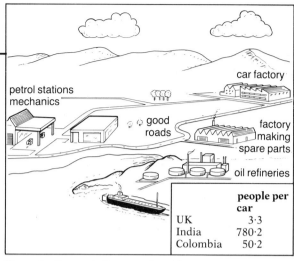

	people per car
UK	3·3
India	780·2
Colombia	50·2

Figure 7   *Cars as indicators of wealth*

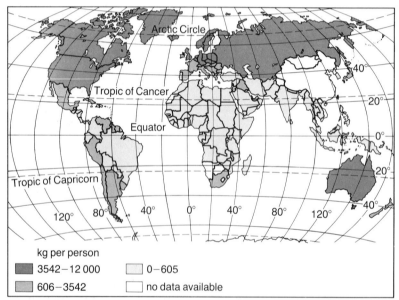

kg per person

■ 3542–12 000	□ 0–605
▨ 606–3542	□ no data available

Figure 8   *World use of energy (energy use is expressed as kg of used per person in a year)*

Using energy

The amount of **energy** used tells about conditions at work and in the home. Electrical energy is used to power machinery. At home it is used for washing machines, heating and computers.

In the poorest countries, people and animals provide most of the power (Figure 9). These types of energy are not shown on the map.

WITH OUR LUCK, THE WORLD'S OIL WILL RUN OUT BEFORE WE CAN AFFORD A CAR!

Figure 9   *Energy and the poor*

Exercises

(Heading: *Possessions and energy*)

7 a Make a drawing like Figure 7 to show how telephones can tell us about the wealth in a country.
   b Choose one of the things listed below. Explain how it could be used as an index of how rich a country is:
      number of television or microcomputer sets
      railway or motorway lengths

8 a What does the term 'energy' mean?
   b Why is measuring energy a good way to tell how rich a country is?
   c What does Figure 8 tell you about the pattern of energy use around the world? (The word pattern means where things are, for example, all together in one area or scattered about everywhere.)

Figures that agree

One index on its own can not show if a country
is rich or poor. But different indices often tell
the same story.

In Figure 11, figures have been plotted to
see if there is any connection between different
indices. There is a **positive correlation** if one set
of figures goes up while the others also go up. In
a **negative correlation**, the opposite happens. A
correlation does not mean that one thing
causes another to happen.

		GNP	% farming	energy per head in kg coal equivalent
1	Australia	9820	6	6500
2	UK	7920	2	5300
3	Italy	6480	14	3000
4	Hungary	4180	22	3700
5	Mexico	2130	40	1300
6	Guatemala	1110	57	240
7	Mali	190	87	25
8	Ghana	420	52	180
9	Peru	930	40	680
10	Liberia	520	72	400

Figure 10   *GNP, farming and energy*

Dividing the world

The world map shows a simple division of the
world's countries into rich and poor (Figure
12). The terms **north** and **south** are sometimes
used to describe these areas. The poor countries
are also called **the third world** or **developing**
countries. Many different indices are used to
make these divisions.

Changing fast

A division into two parts is too simple. There
are about 120 countries in the world. Dividing
them up needs many more than two categories.

Some countries are changing quickly as more
roads, factories and houses are built. They are
said to be becoming more **developed**. In other
countries, **development** is taking place far more
slowly. Even between the poor countries, there
are very great differences in the standards of
living.

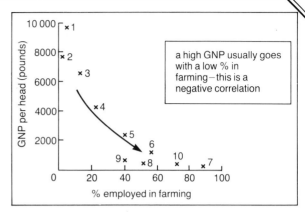

Figure 11   *Farming and GNP*

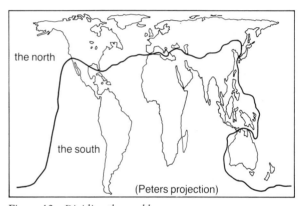

Figure 12   *Dividing the world*

Exercises

(Heading: *North and south*)

9   a Make a copy of Figure 11.
    b Plot figures for GNP against figures for
       energy used per head (Figure 10).
    c Say if there is a positive or a negative
       correlation between the figures.
    d Write a paragraph to explain how one set
       of figures might affect the other.

10 On a world outline map, draw in the
    dividing line between the countries that are
    mainly rich and those that are mainly poor.

11 a What is meant by the term 'developing
       country'?
    b What is wrong with the simple division
       of the world's countries into two groups?

9

1.3 Food and water

About 500 million people in the world go hungry every day. Almost all of them live in the developing countries. Yet there is no world food shortage. About 10% more food is grown every year than is needed. There are mountains of unsold grain and dairy produce in Europe and North America.

Measuring food needs

Figure 2 shows how much food is eaten in different parts of the world. In Europe and North America, overeating is a health problem. In Africa, Asia and South America, most people eat less than 2400 kcal each day. These people suffer from **undernourishment**.

Below the breadline

Figures for a whole continent hide the fact that in many countries, millions of people eat well below the average. At below 1600 kcal per day, there is a danger of death through **starvation**.

In India, 80 million out of 720 million people have less than half their day's basic needs. In Bangladesh, 75% of all children are either partly or severely undernourished.

Diet and disease

Food and disease are directly linked. Severe lack of food causes diseases such as marasmus and kwashiorkor. Both of these can kill.

It is also important to eat the right kinds of food. Without the right balance of protein and energy, there is likely to be **protein – energy malnutrition**(PEM). Lack of some vitamins causes special problems, such as blindness due to lack of vitamin A.

Exercises

(Heading: *Measuring food*)

1 Why is measuring food a good way to tell how poor people are? Look at Figure 1 and say how little else the family has.

Figure 1 *A mother feeds her children. Will there be enough?*

	kcal
North America	3530
Western Europe	3390
Oceania	3370
other developed market economies	2850
Eastern Europe and the USSR	3460
Latin America	2540
Near East	2440
Africa	2110
Far East	2040
other developing market economies	2340
Asian centrally planned economies	2290

0   1600   2400
average requirement

Figure 2 *Daily food energy supplies (kcal)*

2 What do each of these terms mean: malnutrition...2400 kcal per day... undernourishment...PEM

3 a Name some health problems caused by lack of food, or eating an unbalanced diet.
   b Poor health affects how much work can be done. Explain how this adds to a farmer's problems in growing more food.

Growing enough food

There is no one reason why so many people do not have enough to eat. Growing enough food is a problem in many countries. Figure 3 shows some reasons why this can happen.

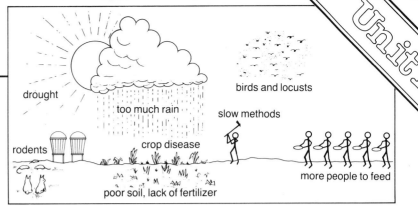

Figure 3 *Problems of growing food*

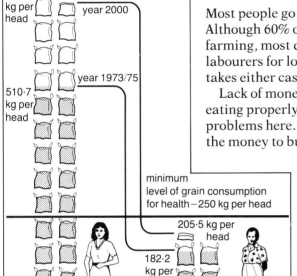

Figure 4 *A widening gap*

Poverty and hunger

Most people go hungry because they cannot afford to buy food. Although 60% of people in developing countries work in farming, most do not own the land they work on. Some work as labourers for low wages. Othes rent land from a landlord who takes either cash or part of the crop as payment.

Lack of money also stops poor people in urban areas from eating properly. High unemployment and low wages are the problems here. Growing more food will not help those without the money to buy it.

The gap widens

People in rich countries play a part in the causes of hunger. Land in poor countries is sometimes used to grow export crops such as sugar and animal feed. This brings some money into the poor country, but the land might be better used in growing more food for the poor. Figure 4 shows how the gap seems to be widening as the rich take a growing share of the world's food.

Exercises

4 Study Figure 3 then list the problems in growing more food in some tropical countries such as Africa or Asia.

5 Give reasons why people in both town and country areas often can not afford to buy food.

6 a How can export crops cause food shortages in poor countries?
   b What does Figure 4 tell you about the amount of food eaten by people in rich and in poor countries? What is likely to happen in the future?
   c Why are the mountains of food that cannot be sold in the rich countries not sent to the poor countries?

Figure 5   *Carrying water in Africa*

The water carriers

Women carrying water is a common sight in developing countries (Figure 5). Piped water to the home is rare, so it has to be carried from wells, rivers and ponds. Even in urban areas, fetching water is part of the daily routine (Figure 6). The job takes several hours a day, and uses up energy.

A health hazard

Eight out of every ten diseases in poor countries are caused by unclean water (Figure 7). Sources of drinking water may also be used for animals, as open sewers for factory waste, and to wash away human waste. In hot climates, dirty water makes ideal breeding grounds for insects and worms.

Water spreads disease from person to person, and from village to village. Solving problems of water supply and sewerage is a key to improving the quality of life for the poor.

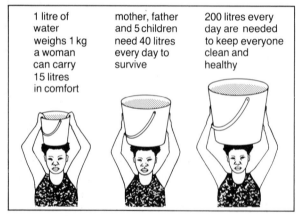

| 1 litre of water weighs 1 kg a woman can carry 15 litres in comfort | mother, father and 5 children need 40 litres every day to survive | 200 litres every day are needed to keep everyone clean and healthy |

Figure 6   *The weight of water*

disease	deaths each year
diarrhoea	5−10 million
malaria	1·2 million
bilharzia	0·5−1 million
hookworm	50−60 thousand
river blindness	20−50 thousand
amoebiasis	30 thousand
roundworm	20 thousand
polio	10−20 thousand
typhoid	25 thousand
sleeping sickness	5 thousand

many survive these illnesses but are weakened by them for the rest of their lives

Figure 7   *Diseases spread by water*

Exercises

(Heading: *Water supplies*)

7   Draw a sketch of Figure 5. Add a speech bubble to say what the woman might say about carrying water. Think about these things:

    meeting friends
    time taken each day
    the weight
    why women do this job

8   a   Draw a diagram with labels to show why water is often unsafe for people in developing countries. The labels should show the different ways rivers or ponds are used.

    b   How does water spread disease?

9   a   What are some of the problems in improving water supplies in developing countries. Here are some clues:
        cost...distances...rainfall
        ...source of water.

    b   Explain why money spent on improving water supplies in poor countries is money well spent.

1.4   Two out of three

The odds of being born and living in a rich developed country are only one in three. Two out of every three people live in the developing countries. In these countries, the chance of a long and happy life is small.

Who lives where

Figure 1 shows where the world's people live. About 1·1 billion live in the developed countries in Europe, North America, Oceania and the USSR. The majority, 3·3 billion, live in the poorer countries in Africa, Asia and Latin America.

Going up

The most important fact about world population figures is that they are increasing very quickly (Figure 2). The term **population explosion** is used to describe this increase.

The second fact is that the population increase is very much faster in the developing countries compared with that in the developed countries.

Time to double

The percentage increase figures for developing countries (Figure 1) may seem low. But this is enough to double the population total in these countries in only 33 years. The increase of 0.4% would take 187 years to double the population of Europe.

The rate of increase has been falling since the 1960s, but it is still high enough to double the world's population in only 40 years.

Exercises

(Heading: *World population*)

1   a   Make a chart to list the continents in order of population size. See Figure 1.
    b   What does the chart tell you about how rich or poor most people in the world are?

area	present population	% increase	doubling time (years)
World	4585	1·7	40
Africa	498	2·9	24
Asia	2671	·9	37
North America	256	0·7	95
Latin America	378	2·3	30
Europe	488	0·4	187
USSR	270	0·8	88
Oceania	24	0·8	88

Figure 1   *World population totals*

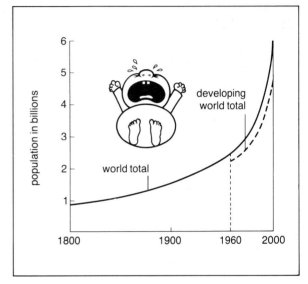

Figure 2   *World population increase*

2   Explain the term 'population explosion'. Use figures from the graph to add detail to your answer. See Figure 2.

3   a   What do you notice about the speed at which the population is increasing in the poor countries of Asia, South America and Africa?
    b   How does the increase in poor countries compare with the increase in the rich continents?
    c   Write a paragraph to outline some of the problems you think there might be when the population of a poor country doubles. Think about housing, hospitals, schools and food.

Births and deaths

Population increase happens when there are more births than deaths. **Birth rate** and **death rate** figures show how many people are being born and how many die for every 1000 people in a year. The difference between them is called the **natural increase**.

The natural increase figure makes it easy to compare one country with another, even though they have different population totals.

Figure 3 shows typical birth and death rates for some countries. The greater the difference between births and deaths, the larger the increase will be (Figure 4).

Living longer

In the past, many children did not survive to become adults. In developing countries today, 100 out of every 1000 children die before their first birthday. This figure is called the **infant mortality rate**. Thirty years ago, the infant mortality rate was twice as high.

People are also living longer (Figures 5 and 6). Over the same 30 years, average **life expectancy** in developing countries increased from 42 to 54 years of age. A low infant mortality rate and a long life expectancy means that more people live to become parents. This increases the population total year by year.

country	birth rate per 1000 people	death rate per 1000 people	infant mortality per 1000 live births
Nigeria	50	18	135
Pakistan	44	16	126
Honduras	47	12	88
UK	14	12	12
Hungary	14	14	23
Canada	16	7	11

Figure 3 *Birth and death rates*

	birth rate	death rate	natural increase per 1000 each year
UK	14	12	2
Nigeria	50	18	32

Figure 4 *Working out natural increase*

country	age
Nigeria	48
Pakistan	51
Honduras	57
UK	73
Hungary	70
Canada	74

Figure 5 *Life expectancy*

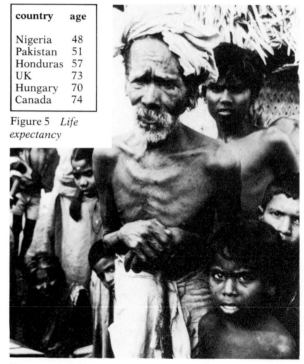

Figure 6 *Old age in an Indian village*

Exercises

(Heading: *Population statistics*)

4 Copy and complete these sentences with a suitable figure as an example:
   a A high birth rate figure is about _____.
   b A high death rate figure is _____.
   c A low birth rate figure is _____.
   d A low death rate figure is _____.
   e Infant mortality in a poor country is about _____.
   f A high rate of natural increase is _____.
   g In a poor country, people can expect to live until the age of _____.

5 Explain how fewer deaths among children and longer life expectancy can lead to a high rate of population increase.

Population pyramids

A **population pyramid** diagram is an easy way to show the **population structure** of a country (Figure 7). This shows the age of people, and the percentage of males and females.

In developing countries, between 40% and 50% of the population is under 15 years old. Those over 65 account for only 5%. Figures 8a and 8b compares population pyramids in developed and developing countries.

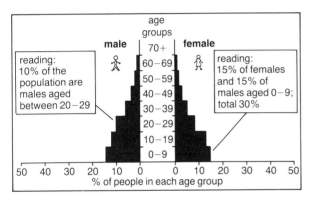

Figure 7 *Reading a population pyramid*

Figure 8 *a Developed-world pyramid* *b Developing-world pyramid*

Figure 9 *Health education about cholera*

Keeping pace

Better medical care has meant fewer child deaths and the chance of a longer life. In some areas, this is helping to lower the birth rate. Parents do not need to have as many children.

A problem now is how to bring medical help to the extra people. This is expensive and needs trained doctors and health workers. Killer diseases such as cholera still exist in spite of attempts to control them (Figure 9).

Exercises

(Heading: *Population pyramids*)

6 a Copy Figures 8a and 8b.
  b For each diagram, add these labelled notes to describe the shapes you have drawn:
    developed country
    low birth rate
    most children survive to become teenagers
    a long life expectancy for most people
    developing country
    a high birth rate with many children
    many children die while still very young
    more deaths as people grow older
    short life expectancy
  c Sketch or describe how the shape of a pyramid in a poor country would change as the birth rate declines and there are fewer deaths among children.

(Heading: *The reasons for population increase*)

7 Study the list below showing some reasons for population change in a country. For each, say how you think it affects population totals both now and in the future:
    new hospitals...good harvest...
    famine...migration...war
    ...people become richer
    ...high unemployment

Figure 10   *Family sizes in rich and poor countries*

	% under 15	% over 65
world	35	6
more developed	23	11
less developed	39	11
Africa	45	3
Asia	36	4
North America	24	9
Latin America	40	4
Europe	23	13
USSR	24	10
Oceania	31	8

Figure 11   *Children and old people*

job	age
minding chickens	8
care of babies	8
fetching water	9
planting rice	10
earning money	13

Figure 12   *Children help to do the work*

Family size

In Figure 10, two families are shown. In the developing country, parents and their children live with grandparents, aunts, uncles, and other relatives. They live in an **extended family**.

Children and old people

In rich countries, bringing up children is expensive. Old people have to be supported with pensions. The numbers of children and old people in a population is called the **dependancy load** (Figure 11).

In poor countries, a large family is needed to share the work. This is most important in the farming areas. Children do useful jobs such as looking after babies and doing household chores while both parents work in the fields (Figure 12).

Nowadays many young people go to work in the cities. They continue to help the family by sending money home.

It's the custom

Sometimes religious beliefs and local customs mean that more children are needed. In India, the family farm is usually handed on to a son. A son may also be needed in funeral ceremonies.

Exercises

(Heading: *Family life in developing countries*)

8   Explain why it is useful to have a large family in a poor country, especially in a rural area. Mention these things:
   care for the old...
   household work...work on
   the farm...earning money

9   Say which person in the family ought to do the following jobs:
   fetching water...cooking...ploughing...
   looking after cattle...making clothes
   ...weeding...going to market...planting
   crops...collecting wood
   Give reasons for your choices.

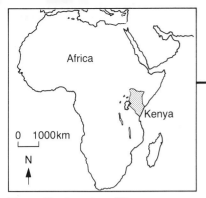

Figure 13   *Location of Kenya*

The people of Kenya

Kenya became an
independant country in 1963.
Since then it has become one
of the more successful
developing countries in
tropical Africa (Figure 13).

The information in Figure
14 gives details of Kenya's
population. Information like
this can be used to plan for the
people's present and future
needs. Housing, health and
jobs all have to be planned for.
The study of population is
called **demography**.

Figure 15   *Mother and child in a Kenyan village*

area	582 646	km²
population	15 865 000	
average density	27	per km²
population % increase	3·5	
total increase per year	597 560	
birth rate	51	per 1000
death rate	13	per 1000
infant (mortality) death rate	87	per 1000
dependancy load %	52	
life expectancy	55	years
population % under 15 years old	50	
population in AD 2000	36 000 000	
average number of children born to women aged 15 – 49	7·6	
urban population %	14	
work force in agriculture %	78	
adult literacy rate %	50	
access to safe water %	17	
people per doctor	14 244	
people per hospital bed	770	
calorie intake as a % of needs	95·6	

Figure 14   *Population statistics for Kenya*

Exercises

(Heading: *The people of Kenya*)

10 a Which figures tell you that the
       population of Kenya is increasing very
       quickly?
    b What is the increase per 1000 each year
       (births minus deaths)?
    c How do the figures show that most
       people in Kenya still live by farming?
    d Explain how the figures show that there
       are problems of poor health in Kenya.

11 a Describe the scene shown in Figure 15.
    b Using the statistics in Figure 14, say
       what you think their present lives are
       like and what the future holds for them.
       Include:
           the rest of the family
           where they live
           food and health
           the work they do

1.5　Poverty belts

The poorest countries on earth stretch in a line through Africa and on into Asia. For the people in these countries, life is a struggle just to survive. Development towards a better life seems a long way off.

	32 LDCs	89 other developing countries	37 developed countries
population in millions	285	3001	1131
infant mortality (per 1000)	160%	94	19
safe water supply	31%	41%	100%
adult literacy rate	28%	55%	98%
GNP per head	$170	$520	$6230
people per doctor	17 000	2700	520

Figure 1　*Statistics for the LDCs*

Life on the margin

The United Nations lists 32 places as being the **least developed countries** (LDCs). They are at the bottom of almost every index used to measure development (Figures 1 and 2).

There are few industries. Farmers grow less food than they need. Most people do not have access to health and education services. Three out of every ten die before the age of 5 and only five out of ten reach the age of 40.

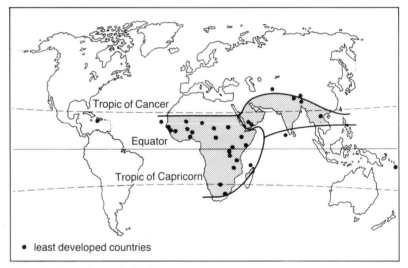

Figure 2　*Location of the LDCs*

Common problems

Many of these countries share the same kind of problems. In Africa, farming in the dry Sahel countries south of the Sahara is difficult, even in an average year. In some years there is **drought**. With no rain, nothing can grow.

The **monsoon** countries of Asia have flooding when the rains are too heavy, and drought when the rains fail to come.

Many of the LDCs are **land-locked**, with no coast. Being land-locked makes importing and exporting more expensive.

In other ways, such as history, government and religion, the LDCs are different from each other.

Exercises

(Heading: *The world's poorest countries*)
1 　a 　On a world outline, shade in the LDCs shown on Figure 2.
　　b 　Use an atlas to name at least 10 LDCs in Africa and 3 in Asia.

2 　What facts tell you that people in the LDCs have problems with:
　　a 　food supply,
　　b 　health?

3 　a 　Explain how the climate of LDCs adds to their problems.
　　b 　Why can development be more difficult for a land-locked country than for one which has access to the sea?

	Burkina Faso	UK
population in millions	6·7	56·1
size in km²	274 000	244 000
natural increase	2.6 %	0·2 %
infant deaths per 1000	211	11·8
life expectancy	42	73
% in farming	82 %	2 %
GNP per person	$190	$7920
urban population	8 %	77 %
people per doctor	57 130	750

Figure 3   *Statistics for Burkina Faso*

Poverty in Burkina Faso *(formerly Upper Volta)*

Burkina Faso is almost bottom of the LDC list. Every index tells the same story of widespread human misery (Figure 3). Average figures for the country do not show the whole truth. Most hospitals and schools are in the towns. The few wage-earning jobs are also there. Conditions in the rural areas are well below the average (Figures 4 and 6).

Exercises

(Heading: *Poverty in Burkina Faso*)

4   a   Use an atlas to draw a sketch map of Burkina Faso to show its main geographical features:
      Burkina Faso and the surrounding countries
      Ougadougou (the capital city)
      relief (hills, lowland or mountain)
      routes to the sea, e.g. railway or road
      the river Volta
      Add a scale line to your map.
    b   Use other atlas maps to add labels to your map about:
      the climate
      the natural vegetation

5   Draw diagrams to illustrate the statistics in Figure 3.

6   Look at Figures 4, 5 and 6. Write a paragraph for each to say what they tell about the way of life of people in Burkina Faso.

Figure 4   *A village 50 km from the capital*

Figure 5   *Off to work in Ougadougou*

Figure 6   *Making cotton yarn by hand*

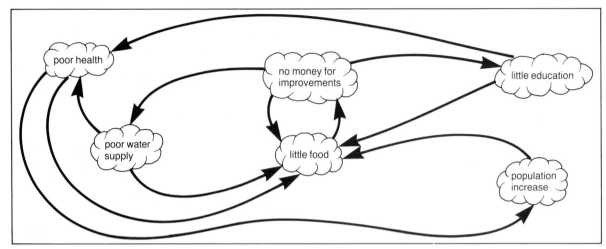

Figure 7   *Poverty links*

Round in circles

It is hard to improve the quality of life for people in Burkina Faso. There are two reasons for this. The first is because problems of health, education, farming and industry are all linked. Figure 7 shows how this happens.

Better water supplies would help grow more food. But nobody will be better off if population increase is not slowed down. The population will continue to increase until there are better health and education services.

Out of control

The second problem is that most people's lives depend on things outside their control. Eight people out of every 10 depend on farming. But farmers can do nothing about the amount of rain that falls. With no money to spare, they cannot afford to build reservoirs or drill wells.

There are few resources for industry so imports such as oil are needed. The price of imports such as oil depends on what other countries charge.

Groundnuts are the main export crop from Burkina Faso, but other Sahel countries do the same. The price for groundnuts goes up and down every year, depending on how much everyone produces.

Where to start

To reduce poverty, there must be a plan which tries to solve all these problems at once. This kind of plan would be very expensive. It is hard to see where the money will come from.

Exercises

(Heading: *Poverty circles*)

7   Study Figure 7. Use examples to explain how the problems holding back development are linked. Explain for example, how education and health are linked.

8   a Why are farmers not able to grow more food?
    b What are the problems of relying on other countries for items such as oil?
    c Explain why the price for Burkina Faso's export crops changes so much year by year.

9   a Why is it no use trying to solve the problems of Burkina Faso one at a time?
    b Why is it so hard for the government in Burkina Faso to get enough money to solve all the problems?

Figure 8

Countries in collapse

Problems of poverty seem to be affecting more and more people. Droughts keep life on a subsistence level (Figures 8 and 9). There are no opportunities to save money so that farming can be improved.

The effects spread beyond the boundaries of each country. Poverty and hunger lead to unrest, sometimes to wars. Loans can not be repaid and trade in foods, raw materials and goods is affected. Problems on this scale have effects in many countries, rich and poor, around the globe.

AFRICA FACES CATASTROPHE

Observer 18 March 1984

Several African countries suffering from famine are near to catastrophe. World Bank experts say that unless a much more vigorous aid programme is undertaken, these countries will collapse entirely and go back to being 'bush' economies.

An internal report is believed to give four of them – Chad, Ethiopia, Upper Volta and Mocambique – only four months at present relief rates, before disaster. Mauritania, Ghana, Mali and Senegal are also in immediate danger.

Fourteen other African countries are severely hit by the drought, or rather by a series of droughts, which in the word of one expert threaten a 'continental' disaster. The first eight are totally unable to pay for food imports.

Even where aid is arriving, it is often impossible to get it to stricken areas for want of roads, trucks and effective administration. The UN Food and Agriculture Organization estimates that 20 million now face starvation. Tens of thousands have already died.

Poor rains, are only one cause of the crisis. Others include cattle disease (rinderpest), the spread of a crop-eating insect (the grain-borer), and the breakdown of security, as in

Chad, Ethiopia and Mocambique.

There are also long-term causes. Another report now ringing alarm bells for Britain, the Commonwealth, the United Nations and the European Community – all of which have major interests or responsibilities in Africa – speaks of a 20-year decline in *per capita* agricultural production.

This is due particularly to mistaken economic policies favouring cheap food for city dwellers at the expense of farmers who, as a result have left the land and gone to live in the cities.

Collapse throughout the continent would have disastrous consequences for world health, world trade, and international security.

Figure 9   *A newspaper report*

Exercises

(Heading: *Countries in collapse*)

10 On an outline map of Africa, shade in and name the countries listed in Figure 9.

11 a What climatic problem is causing the famine described in Figure 9?
    b Why will the inability to grow food specially affect so many people in a developing country? (*Hint*: employment structure)
    c Why can food not be bought from other countries?

12 a What problems, apart from climate, are causing problems to growing more food?
    b Security means being safe from war. How can this affect food supply?
    c How have the policies of rich countries caused less food to be grown in many African countries?

13 In what ways can famine on such a scale as described in Figure 9, affect people in the rich countries?

1.6 Turkey old and new

Turkey is south of the North-South line between rich and poor countries. It is a developing country, but is as different from the poorest such as Burkina Faso, as it is from rich developed countries such as the UK.

Turkey photofile

Figures 1, 2, 3, 4 and 5 show scenes from different parts of Turkey. Some of the scenes have not changed much over hundreds of years. Others show that Turkey is becoming a modern developed country.

Figure 3　*Mechanized farming in Turkey*

Figure 1　*Istanbul city*

Figure 4　*Traditional farming methods*

Figure 2　*Village life in Turkey*

Figure 5　*Modern transport in Turkey*

	cultivated land in '000 hectares	land cultivated by tractors	work			labour force in farming	
			unemployment rate		18%	1960	78%
1960	23 260	3160	work force in	– agriculture	56%	1970	71%
1970	24 300	7940		– industry	11%	1980	56%
1980	24 750	23 500		– services	33%		

electrical energy production in millions of kilowatt hours		income per head		GNP in billion TL – Turkish currency (1968)	
1975	15 620	1975	$885	1970	147·7
1980	23 290	1980	$1114	1980	4484·3

Figure 6   *Economic statistics for Turkey*

Catching up

Turkey is trying to catch up with the **industrialized** countries such as the UK, USA and Japan. Figure 6 shows some of the ways that the economy is becoming more developed.

Friends and neighbours

Links with other countries are important in building up a modern economy. Turkey has special **trade** agreements with the Common Market (EEC) countries.

Turkey is also an important member of the NATO defence treaty. Turkey's northern borders are shared with the USSR and other East European communist countries.

At the same time, Turkey is an Islamic country. In culture and customs, it has strong links with other Arab countries in the Middle East.

Exercises

(Heading: *Turkey old and new*)
1   What do the photographs show about each of the following:
   a   modern city life;
   b   village work and houses;
   c   new farming methods;
   d   old farm methods;
   e   transport?

2   Do you think that you get an accurate picture of Turkey by looking at Figures 1, 2, 3, 4 and 5? Are there any other scenes you would need to see to get a complete picture?

(Heading: *Development in Turkey*)
3   a   In what way do the figures show that farming is becoming more modern?
   b   What do the figures for electricity production tell you about Turkish development?
   c   Average incomes have been rising steadily. What was the increase between 1975 and 1980?
   d   How did the GNP figures change between 1970 and 1980. What could have caused the change?

4   Turkey is sometimes included in maps of Europe. On other maps, it is drawn as part of Asia. Explain why this is so.

	population total in millions	% urban		birth rate	death rate	life expectancy		
1945	18·8	25	1955 – 60	45	18	% increase for 1982	2·2	%
1955	24·1	29	1960 – 65	41	15	life expectancy	60	years
1965	31·4	34	1965 – 70	41	13	population under 15		
1975	40·3	42	1970 – 75	37	12	years old	40	%
1980	42·2	44	1975 – 80	37	11			
			1982	30	10			

Figure 7   *Population statistics for Turkey*

A changing population

Population figures have changed quickly over the last 20 years (Figure 7). At the moment, Turkey has the same kind of population problems as other developing countries. There is a high population increase and more people are moving to the cities. However, Turkey is one of the richer developing countries. This helps it cope better with these problems (Figure 8).

Figure 8   *New housing in Turkey*

education	adult literacy	population per doctor	
1960	40%	1960	3000
1975	55%	1975	2130
		1980	1700
access to safe water		**possessions**	
1975	68%	people per car	47
		people per telephone	31

Figure 9   *The standard of living in Turkey*

Life for the people

Information about the standard of living for the people is given in Figure 9. Remember that these figures do not show differences around the country.

Much of central Turkey is mountainous and difficult to get to. Life in these parts is very different from life in a modern city such as Istanbul.

Exercises

(Heading: *The people of Turkey*)

5   Use statistics from Figure 7 to show how Turkey's population is typical of developing countries in each of these ways:
   a   there is a fast population increase;
   b   there is a high birth rate and falling death rate;
   c   a large percentage are under 15 years old;
   d   an increasing percentage live in urban areas.

6   Describe the new housing scheme shown in Figure 8. Mention these things:
   the relief of the land
   the street layout and house spacing
   the style and building materials of the houses
   who you think will be able to afford the houses

7   Use the information in Figure 9 to say if living standards in Turkey have been improving in recent years.

Unit 2   Brakes on progress

2.1 Food for the world

Poor farmers will have to grow 50% more food if their families are to have enough food to eat in the year AD 2000.

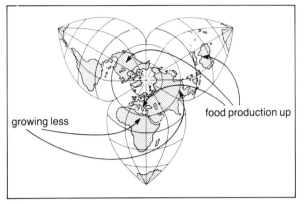

Figure 1   *World food report*

Growing more

The world can grow enough food to feed itself. There are still vast new areas to be farmed. Africa for instance can feed three times its present population. More irrigation and fertilizers are major ways of increasing the world's food supply.

But the poor will not be able to grow enough unless they have a larger share of the land. Governments and landowners will have to provide them with much more land.

Exercises

(Heading: *Food for the world*)

1   Study Figure 1.
   a Name two continents where less food is being grown.
   b What type of country grew more food?
   c What happened in the poorest countries?

2   Match these descriptions with the correct photos, in Figure 2a, b, c, d, e. Give your list a title.
      watering the land
      more food with better equipment
      more food by terracing the land
      better varieties of seeds
      protecting from pests and diseases

Figure 2   *a*   *b*

c   *d*   *e*

2.2　The land provides

Farming in the Third World is seldom easy. Farming families must be skilful if they are to survive.

A dry land

The Guidimaka region is in southern Mauritania in Africa. It is a **semi-arid** area where little rain falls. Trees and grasses called **savanna** grow in scattered patches.

The area is thinly peopled with only 5 to 10 people per km². The total population is only 50 000, mostly living in villages near the rivers (Figure 1). They grow crops to feed themselves. People who do this are called **subsistence farmers**.

Using what's around you

Farmers in Guidimaka make the best use they can of the climate, soils and water supply.

There is a wide variety of soils and drainage conditions. Farmers tailor their crops and farming methods to suit the conditions (Figure 2).

Water from rain and flood

Land near the rivers is flooded in summer. Mud and silt called **alluvium** is left after the floods. This gives a very fertile soil. As the flood waters fall, the farmers plant their crops. This is called **flood-retreat** farming.

In contrast, land which is not flooded relies on the summer rains. **Rain-fed** or **dryland farming** is necessary. In these places, the soil is sandy and infertile. Farmers seek out the best sites for their plots. Hollows and the lower slopes are chosen as more moisture collects there.

One problem is that a farmer's plots become scattered, sometimes several kilometres apart.

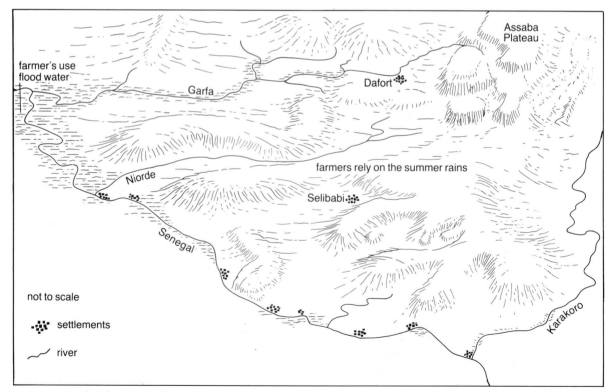

Figure 1　*Guidimaka*

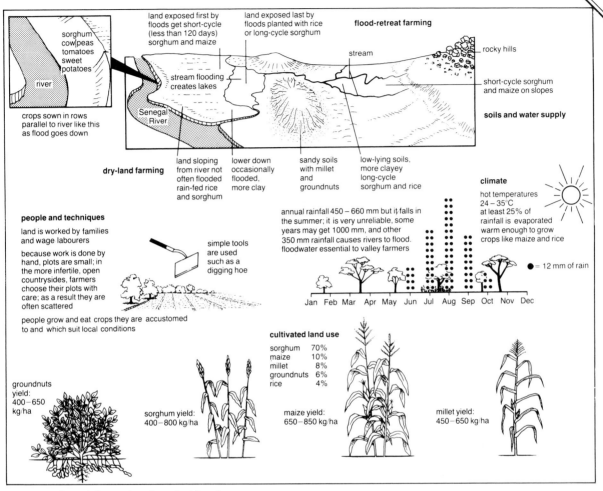

Figure 2 *Some information about Guidimaka*

Exercises

(Heading: *Using what's around you*)

1 What are subsistence farmers?

2 Explain the difference between flood-retreat farming and rain-fed farming.

3 Study Figure 2.

  a List the four main crops grown.
    Which crop takes up most land?
    Which crop gives most food per hectare?

  b Draw a sketch to show how farmers grow different crops to suit different soils and amounts of water. Label on notes about:
    short- and long-cycle crops
    flooded and rain-fed land
    clay and sandy soils
    hollows and slopes

  c How does the climate affect what farmers can do? Mention:
    temperature...type of crops...when rain falls.

  d What other things affect how families farm the land?

4 From what you have read, list evidence to show that families are skilful farmers.

Looking after the land

Farmers know what their crops and soil need. In the rain-fed areas, land is uncultivated for years. This lets the soil regain some fertility. This is called **fallowing**. Stalks from the harvest are left to rot and are dug in for compost. This is the reason for the untidy looking plots (Figure 3).

Against the odds

Growing enough food is always a struggle. Farmwork is done by hand using simple tools. This means that only a limited amount of land can be cultivated by each family (Figure 4). Problems are caused by unreliable amounts of rain and infertile soils (Figure 5). Travel between the plots wastes time and energy.

There is also a labour shortage. In some villages, one third of the young men have left to work in France. They send money back to try to help their family at home. The money is used to help improve their homes. Unfortunately, money is also spent on imported foods rather than growing more of their own.

Exercise

5 Study Figures 4 and 6.
a State the problem shown by each part of Figure 6.
b Make sketches and notes to show the problems which face farmers in Guidimaka.

Figure 3 *Some land is left fallow*

Figure 4 *Simple tools are used for farmwork*

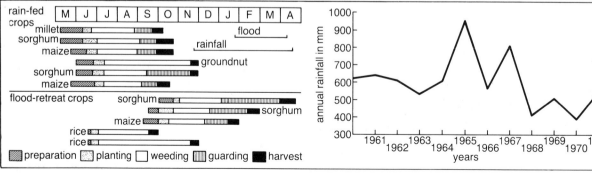

Figure 6 *a Growth of crops*

b Changes in annual rainfall amounts

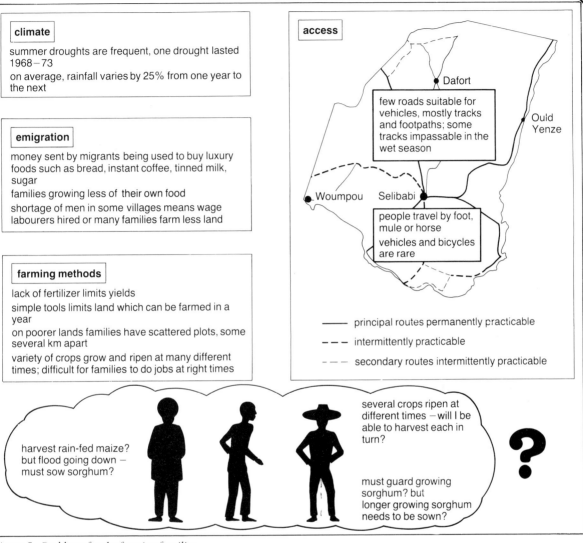

climate

summer droughts are frequent, one drought lasted 1968–73

on average, rainfall varies by 25% from one year to the next

emigration

money sent by migrants being used to buy luxury foods such as bread, instant coffee, tinned milk, sugar

families growing less of their own food

shortage of men in some villages means wage labourers hired or many families farm less land

farming methods

lack of fertilizer limits yields

simple tools limits land which can be farmed in a year

on poorer lands families have scattered plots, some several km apart

variety of crops grow and ripen at many different times; difficult for families to do jobs at right times

access

few roads suitable for vehicles, mostly tracks and footpaths; some tracks impassable in the wet season

people travel by foot, mule or horse

vehicles and bicycles are rare

Dafort

Ould Yenze

Woumpou   Selibabi

——— principal routes permanently practicable

– – – intermittently practicable

- - - secondary routes intermittently practicable

harvest rain-fed maize? but flood going down – must sow sorghum?

several crops ripen at different times – will I be able to harvest each in turn?

must guard growing sorghum? but longer growing sorghum needs to be sown?

Figure 5 *Problems for the farming families*

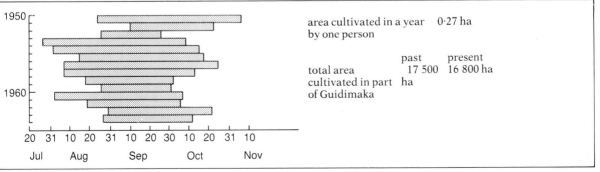

area cultivated in a year   0·27 ha
by one person

	past	present
total area cultivated in part of Guidimaka	17 500 ha	16 800 ha

c Duration of active flooding of the Senegal River at Bakel   *d Area cultivated*

29

Unit 2.3 Disasters and disease

Between 1947 and 1980, over 1 million people died in disasters caused by floods, earthquakes and hurricanes. A further 25 million died in wars. These figures are dwarfed by the numbers who die or are disabled by diseases.

Natural disasters

Disasters caused by floods, earthquakes and hurricanes are called **natural disasters**. Most deaths from natural disasters are in Third World countries (Figure 1).

One example in 1980 was an earthquake at El Asnam in Algeria. This caused 20 000 deaths and made 30 000 people homeless. Most of the damage occurred in just a few hours.

Help needed

Conditions in poor countries are such that disasters are more widespread and cause more destruction (Figure 2). Rich countries can take measures to lessen the effects of disasters.

Countries such as Algeria can only cope if help comes from the richer countries. Emergency help with blankets, tents and medicines is needed. Longer term help is also needed to design stronger buildings and set up better health care.

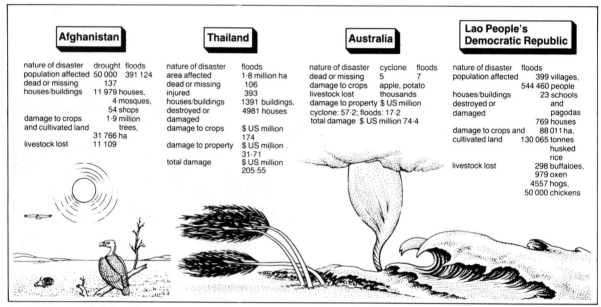

Afghanistan

nature of disaster	drought floods
population affected	50 000 391 124
dead or missing	137
houses/buildings	11 979 houses,
	4 mosques,
	54 shops
damage to crops	1·9 million
and cultivated land	trees,
	31 766 ha
livestock lost	11 109

Thailand

nature of disaster	floods
area affected	1·8 million ha
dead or missing	106
injured	393
houses/buildings	1391 buildings,
destroyed or	4981 houses
damaged	
damage to crops	$ US million
	174
damage to property	$ US million .
	31·71
total damage	$ US million
	205·55

Australia

nature of disaster	cyclone floods
dead or missing	5 7
damage to crops	apple, potato
livestock lost	thousands
damage to property	$ US million
cyclone: 57·2; floods: 17·2	
total damage $ US million 74·4	

Lao People's Democratic Republic

nature of disaster	floods
population affected	399 villages,
	544 460 people
houses/buildings	23 schools
destroyed or	and
damaged	pagodas
	769 houses
damage to crops and	88 011 ha,
cultivated land	130 065 tonnes
	husked
	rice
livestock lost	298 buffaloes,
	979 oxen
	4557 hogs,
	50 000 chickens

Figure 1   *Natural disasters*

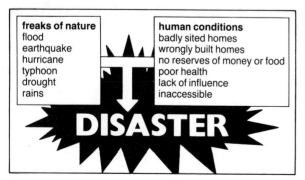

freaks of nature	human conditions
flood	badly sited homes
earthquake	wrongly built homes
hurricane	no reserves of money or food
typhoon	poor health
drought	lack of influence
rains	inaccessible

DISASTER

Figure 2   *How, why and where do disasters strike?*

Exercises

(Heading: *Disasters*)

1   Choose one of the poor countries in Figure 1. Design a front page for a newspaper which reported the disaster. Think up some headlines and short reports of what happened to families.

2   Choose another poor country and contrast what happened with the disasters in Australia.

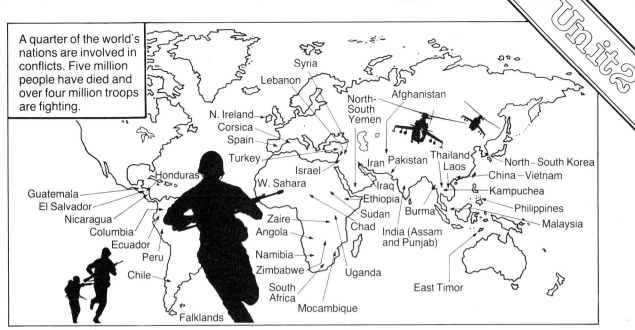

> A quarter of the world's nations are involved in conflicts. Five million people have died and over four million troops are fighting.

Figure 3   *The world at war*

The world at war

Each year, several major conflicts are taking place, most of them in Third World countries (Figure 3). Wars are **man-made disasters**. They bring death and destruction to countries where people already have to struggle to survive.

The tragedy of Vietnam

From 1965 to 1975, US troops supported the South Vietnam army against North Vietnam and south Vietnamese guerillas. The cost to both sides was heavy. Nine million people fled their homes and became **refugees**. Nearly 3 million people were killed.

Bombing caused most of the destruction (Figure 4). Three times more bombs were dropped by US planes than fell on Italy, Japan and Germany in World War 2. People in the new country of Vietnam have been left to rebuild their country.

Figure 4   *The war in Vietnam*

Exercises

(Heading: *The world at war*)

3   a   Copy Figure 3.
   b   Underline the developing countries. What do you notice?
   c   Choose one of these countries and find out why there is fighting.

4   Figure 4 shows some of the effects of the Vietnam War. Describe how the war would have affected the lives of farmers and their families.

Health in the Sudan

The Gezira and Rahad area in the Sudan covers an area of 1 million hectares. A population of 2 million people live there (Figure 5). Cotton is the main crop. It makes up 60% of Sudan's exports.

Compared to other parts of the country, it is a prosperous region. Farmers earn six times more than most other Sudanese farmers.

Irrigation and disease

Irrigation has played a major part in making the land fertile. Unfortunately, the irrigation canals and ditches have also caused health problems. Bilharzia, malaria and diarrhoeal diseases flourish in the warm water. Snails carrying the bilharzia parasite live in the water. People get diarrhoea because they use infected water.

Diseases are widespread. Bilharzia affects 50–70% of people in Gezira. In some villages in Rahad, 60% of the children under 4 years old have diarrhoea. About half of them will die because of it (Figure 6).

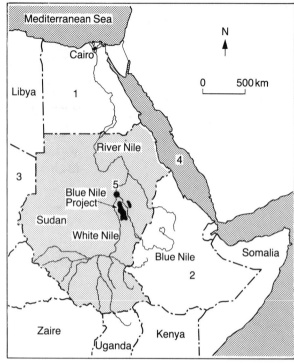

Figure 5 *Sudan*

Exercises

(Heading: *Irrigation and disease*)
5   a Make a copy of Figure 5.
     b Name the following:
         the countries 1, 2 and 3
         the sea 4
         the city 5
     c Give your map a suitable title.

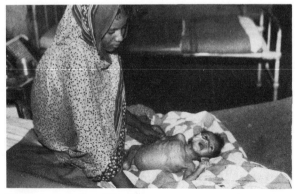

Figure 6 *A victim of diarrhoea*

6   a Name the three main diseases in the Gezira and Rahad area (Blue Nile project area in Figure 5).
     b Write down two facts from the text which show how serious the diseases are.

The Blue Nile project

The Blue Nile health project is unique. For the first time in the world, all the diseases and health problems in an area are being tackled at the same time. The project has several aims. For this reason, it is called a **multi-purpose scheme**.

The diseases are being attacked on several fronts (Figure 7). There is health education for the people. This helps to change conditions which cause disease to spread.

The project will last from 1979 to 1990. The cost of $155 million is high, but the Sudan is being helped out by the World Health Organization (WHO) and countries such as Japan.

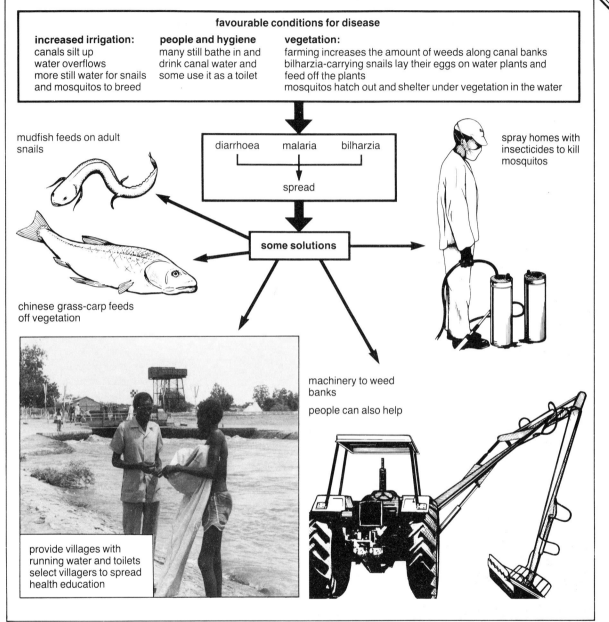

favourable conditions for disease

increased irrigation:
canals silt up
water overflows
more still water for snails
and mosquitos to breed

people and hygiene
many still bathe in and
drink canal water and
some use it as a toilet

vegetation:
farming increases the amount of weeds along canal banks
bilharzia-carrying snails lay their eggs on water plants and
feed off the plants
mosquitos hatch out and shelter under vegetation in the water

mudfish feeds on adult
snails

diarrhoea   malaria   bilharzia

spread

spray homes with
insecticides to kill
mosquitos

chinese grass-carp feeds
off vegetation

some solutions

machinery to weed
banks

people can also help

provide villages with
running water and toilets
select villagers to spread
health education

Figure 7   *The Blue Nile Project*

Exercises

(Heading: *The Blue Nile project*)

7 a In what ways is the Blue Nile project
      unique?
   b Why is it called a multi-purpose project?

8 Study Figure 7.
   a Describe the three main conditions
      which lead to disease.
   b Use sketches and notes to show how the
      diseases are being controlled.

The terrible killer

The historian Macauley called smallpox 'the most terrible of all the ministers of death'. Throughout history, smallpox has been a feared disease. It kills and disfigures. Lumps and scabs cover the body. When these go, scars or pockmarks are left (Figure 8). Smallpox has often played an important part in history (Figure 9). Millions have died and in 1967 alone, the number of deaths was 2 million.

Figure 8 *Smallpox*

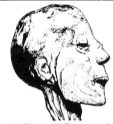

the Pharoah Rameses V died in 1157 BC probably of smallpox

1519 Spanish soldiers carried smallpox to Mexico; as a result over 3 million people died and the Aztec empire was destroyed	1694 outbreak in England; Queen Mary II died of the disease, aged 32	1870 war between France and Germany; smallpox broke out, 23 400 French soldiers died	1978 one woman died in Birmingham when the disease escaped from a laboratory

Figure 9 *Smallpox in history*

Smallpox is dead

Both rich and poor countries worked together to wipe out the disease. In 1966, a **mass vaccination** programme began and outbreaks of the disease started to decline (Figure 10). The last recorded case in the Third World was a hospital cook in Somalia in 1977. Smallpox has been successfully **eradicated**.

Exercises

(Heading: *Smallpox is dead*)

9   State two facts from Figure 9 to show that smallpox was an important disease.

10   Study Figure 10.
 Copy out the correct three statements:
 Smallpox was eradicated in all countries at the same time.
 Rich countries eradicated the disease first.
 Since the 1920s smallpox has steadily declined.
 Both rich and poor countries suffered from the disease.
 Some years saw a sudden increase in the number of people with smallpox.

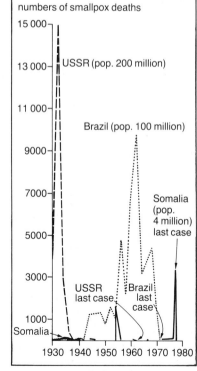

Figure 10 *The decline of smallpox*

child care	providing	fighting	using	making sure	providing a clean water
over half a million mothers die in childbirth	enough food and a balanced diet	diseases	traditional healers and medicines	only essential drugs are used	supply and basic sanitation

Figure 11   *Primary health care*

doctors cost 1000 times more to train

they know about herbal medicines

local villagers decide who should become a health worker

villages help pay wages of about £20 a month; poor people cannot afford expensive doctors

oxen laden with grass matting for shelter, food and two big chests of medicines

donkeys, camels, oxen and even walking can reach remote villages

Figure 12   *Sudanese health worker*

Grassroots medicine

Schemes like the Blue Nile project are expensive and hard to organize. Poor countries have to find easier ways to improve their people's health.

Basic medicines need to be provided. People can be taught and helped to take better care of themselves. In these ways, it is hoped to stop disease breaking out and spreading. Providing for basic health needs is called **primary health care** (Figure 11).

Treatment and tradition

Doctors, health clinics and hospitals are beyond the reach of most people in country areas. As a result, villagers are being trained as **health workers** (Figure 12). They teach primary health care ideas such as hygiene and balanced diets.

Traditional birth attendants are local village women who have helped deliver babies (Figure 13). Because health workers and birth attendants are local people, they are trusted by their communities. People are more willing to learn from people they know and trust.

Figure 13   *A traditional birth attendant*

Exercises

(Heading: *Grassroots medicine*)

11 Look at Figure 11. What kinds of things can poor people learn and do to keep healthy?

12 What is primary health care?

13 Study Figure 12. Why are such workers so suitable to a developing country?

14 a What is the work done by traditional birth attendants?
    b Why are they chosen to teach villagers about health?

2.4 Getting about

Figure 1 *Children carry heavy loads*

About 70% of a farmer's work in poor countries involves lifting and carrying goods. In Nepal, for instance, collecting firewood takes 6 to 8 hours each day.

Lightening the load

There are few roads in rural areas. Most people walk and use tracks and rough paths and have to carry their loads (Figure 1). Bikes, carts and even wheelbarrows make the carrying of loads easier. Sometimes rafts come in useful (Figure 2)!

Too often, however, governments spend money building roads which few people can use. Simple types of transport such as the oxtrike would be more effective (Figure 3). Most farmers need to transport small loads of about 12 kg over short distances of under 20 km.

Figure 2 *Transporting the Third World way*

can be fitted with a small petrol engine

can be made by local craftsmen

'chopper' type wheels to bear heavy loads

other designs possible – for carrying people, drums

good brakes on all wheels

3-speed gear for starting and climbing with heavy loads

Figure 3 *The oxtrike*

Exercise

(Heading: *Lightening the load*)

1 a Name the types of transport in Figure 2.

   b Explain why they are better than using bare hands.

   c Draw a labelled sketch of the oxtrike in Figure 3. Label the advantages it has for carrying goods.

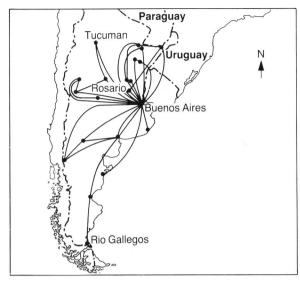

Figure 4   *Airline routes in Argentina*

First things first

Stretches of dual carriageway between cities, or a map of airline routes look impressive (Figure 4). But for the poor, building a track can be more important (Figure 5).

In the Gambia, a 2·5 km path or causeway is being built. It will link the village of Sukuta with its rice fields (Figure 6). It will be usable in all weathers. The path will be made of earth and once it is completed, villagers will not have to wade across streams or through mud (Figure 7). Time and energy will be saved so that more rice land can be cultivated.

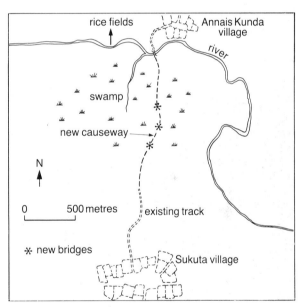

Figure 6   *A dry route across swamp*

Figure 5   *Building a new road*

Figure 7   *The new track*

Exercises

(Heading: *First things first*)

2   a   Make a copy of the map in Figure 6. Give it a title.
    b   Label it with the advantages it will bring to the villagers of Sukuta.

3   Why is the building of tracks or simple roads more important to most people in a poor country?

Figure 8 *A village bridge in Iraq*

Figure 9 *A periodic market in Bolivia*

size	half a hectare
	60 temporary stalls
average number of customers	about 900 on market days
market day	Monday
location	along a road near a cockfighting pit
nearest market	12 kilometres
goods traded (in order of importance)	*rice*, *sugar*, *fish*, groceries, root crops, fruits, vegetables, coconut, livestock, *clothes*, *school equipment*
goods in italics come from outside the area	
how people get there	90% walk
	10% by bus or trucks
cost of transporting goods	farmers living near roads have to pay 5 times less than isolated farmers

Figure 10 *A periodic market in the Philippines*

Far from roads

Millions of people live isolated lives. They rarely visit a town. Distances may be too great or the land mountainous. More often it is the lack of roads and transport which limits how far people can travel (Figure 8). In part of the Philippines, 77% of the population can only be reached by tracks.

Getting to market

People living in remote places have many disadvantages. It can take hours or days to take a sick person to a medical centre. Farmers have to sell their produce at local markets where prices are low. Some goods which are needed such as fertilizers will not be available.

Such markets are small and take place at a convenient spot chosen by the local people. They are set up on one or two days a week and are called **periodic markets** (Figure 9).

When transport is available in remote areas, it is expensive (Figure 10). In contrast, farmers who live near roads have much lower transport costs. They can sell their produce cheaper in the markets.

Exercises

(Heading: *Getting to market*)

4 Describe the difficulties of travel in the area shown on Figure 8.

5 Study Figure 10.
   a Name three goods that local people need.
   b Which are the three main products they sell?
   c Where is the market located?
   d How do most people get there?
   e Why do you think local people do not use the next market town?
   f Why is there more chance of an isolated farmer being poor?
   g Why is periodic market a good name?

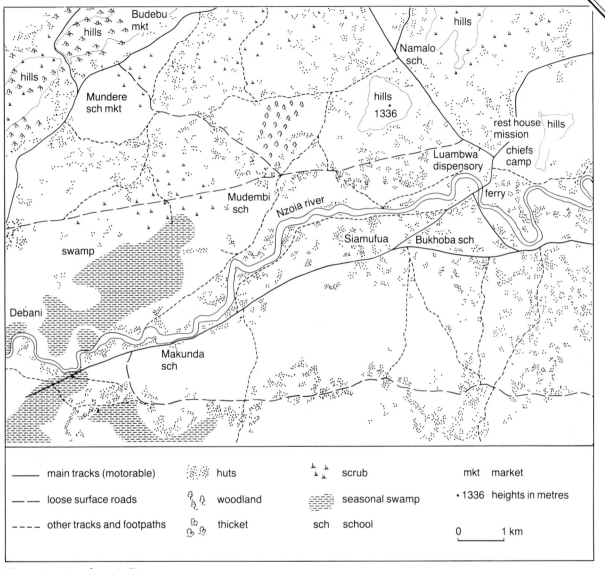

Figure 11   *A rural area in Kenya*

Legend:
- —— main tracks (motorable)
- – – loose surface roads
- - - - - other tracks and footpaths
- huts
- woodland
- thicket
- scrub
- seasonal swamp
- sch   school
- mkt   market
- • 1336   heights in metres
- 0 ___ 1 km

Exercises

(Heading: *Travel in the Kenyan countryside*)

6   Study Figure 11.

   a   Describe the ways in which the landscape is different from the countryside where you live.

   b   List three types of route.

   c   Which type goes near most people's homes?

   d   What does this tell you about how people get about?

   e   Name three obstacles which are avoided by the routes.

   f   Imagine that you lived in Mudembi. Describe the distances and route you would take to:
   shops...school...market.

   g   What does your answer tell you about the time and energy it takes to go places.

2.5 Learning for living

In 1980, one in four people in the world over 15 years old could not read or write. By the year AD 2000, about 950 million people will still be illiterate, in spite of improvements that are being made.

Figure 1 *A village school*

Keen to learn

Over half the children in the Third World go to primary schools, but many leave before they are 11. Their families need them to work.

Fewer than half will go on to a secondary school. The competition for places is fierce. At one secondary school in Zimbabwe, there were 2000 applicants for only 80 first year places.

Most schools in country areas are badly equipped (Figure 1). Many of the teachers have only basic education. Despite these difficulties, children are keen to learn. Being able to read and write is vital if there is to be any chance of a well paid job.

What to learn

In many poor countries, much of what is taught in secondary schools does not meet the needs of the children. Even passing exams may not lead to a job. Unemployment is much higher than in developed countries, and job opportunities are more limited.

Many children will become farmers so farming is becoming an important subject at schools in some countries.

Exercises

(Heading: *Keen to learn*)

1 a State two difficulties of education in developing countries.
  b Despite these difficulties, why are children so keen to learn?

2 a Why do you think farming is such an important part of school work (Figure 2)?
  b Apart from farming, in what other ways is the timetable different from your own?
  c Give reasons for these differences.

time	Monday	Wednesday	Friday
8.00–8.45 8.45–9.30 9.30–10.00	calculations French practical work	calculations French grammar Kirundi (composition)	calculations French exercises agriculture (practical)
10.00–10.15			
10.15–11.00	practical	environmental studies	agriculture (practical)
11.00–11.45	work	religion	
11.45–14.15			
14.15–15.00	environmental studies	geography	Kirundi (reading)
15.00–15.45	essay writing	hygiene	physical education
15.45–16.00			
16.00–16.30	Kirundi (language)	French (conjugation)	French (exercises)

Figure 2 *Three days at school*

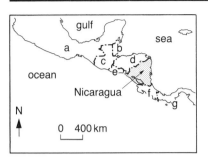

Figure 3 *Central America*

Death to illiteracy

In 1979, there was a revolution in Nicaragua which overthrew the government (Figure 3). The victorious rebels declared 1980 to be the year of literacy. Their new government started the Nicaragua Literacy Crusade.

At the start, Nicaragua had one of the highest illiteracy rates in Latin America. After 6 months, nearly half a million people had been taught to read. About 88% of the population over 10 years old are now literate.

A weapon for change

Everyone helped the Crusade in some way. Those who could read and write helped those who could not (Figure 4). People were taught in attractive and interesting ways (Figure 5).

Literacy brings many benefits and helps a country to become developed. The cartoon about health shows how basic information can be spread once people can read (Figure 6).

Figure 4 *Children as teachers*

Figure 5 *Learning new words*

Figure 6 *Health education*

Exercises

(Heading: *The Nicaragua Literacy Crusade*)

3 a Make a simple sketch of Figure 3.
   b Use an atlas to name the places marked.

4 a What could people learn from Figure 6?
   b Why is this a good way of teaching people about health and hygiene?
   c Try drawing a cartoon about a health problem. Make sure it has a message.

5 Copy out this list of statements. Opposite each statement write evidence to show it is true. Use the text and figures to find the evidence.
   the Literacy Crusade was a success
   large numbers of people were made literate in a short time
   children played an important part

Million dollar men

Agricultural education is needed in Third World countries because so many people depend on farming. Experts from developed countries can help but in the past, their advice has not been suitable. One reason is that they have failed to talk to local farmers and so do not understand all the problems (Figure 7).

Trained workers called **agricultural extension workers** are now being used. They go to country areas and spread knowledge and new ideas. Their task is difficult (Figure 8).

Help yourself

Attitudes are changing (Figure 9). Extension workers are taught to listen and learn from local farmers. They discuss what can be done.

Poor governments do not have enough money to help all the farmers. As a result, people must be encouraged to help each other. When people do things for themselves, it is called **self-help**.

Figure 8　*An agricultural extension worker in Ethiopia*

Figure 9　*An expert ready to listen*

Figure 7　*The million dollar man*

Exercises

(Heading: *Million dollar men*)

6　Why do you think the advice given by the expert in Figure 7 will not be very useful?

7　a　Describe how you think the expert in Figure 9 acts when he comes to a village?
　　b　Why are the villagers more likely to take his advice?

Passing on information

It is impossible for extension workers to visit every village. Instead, people come from a wide area to meetings. Radios, filmstrips and posters are used to pass on information (Figure 10). At such meetings, people can discuss and argue about what they have heard.

Through meetings and radio broadcasts, new ideas reach thousands of farming families. People hear about what is happening in other parts of their country. This helps them feel less isolated. They hear about ideas that have worked for other people (Figure 11).

Exercise

(Heading: *Passing on information*)

8 a What are the advantages of radio and filmstrips for spreading information?

   b Choose one of the descriptions in Figure 11. Describe what is happening.
      Mention:
         why things had to change
         how people got information
         how things improved

Figure 10   *Afghan farmer listening to a tape*

Disease threatened the cassava crop in the People's Republic of the Congo. Radio broadcasted a warning. Crop had to be harvested immediately. Harvesting is normally women's work. Radio suggested men should help. For the first time in memory they did. Cassava crop was saved.	In Tunisia cattle fattened on hay. Was risky because hay could be ruined by rain. Cattle not properly fed. Farmers better off if they fed their cattle with silage. Silage is made from green grass. Less likely to be ruined by the weather. More nutritious. Filmstrips about silage shown to farmers in barns and stables. White-washed walls used as screens. Projectors run off car batteries. Many farmers were won over and began to use silage.	The Government of Tanzania was worried about the people's lack of knowledge about hygiene and sanitation. Nearly a 100 000 people were trained to lead groups in villages. Groups listened to radio broadcasts about health. Discussed what could be done to improve health in their villages. As a result of broadcasts villagers built latrines, cleaned water supplies, destroyed breeding places of mosquitoes which spread malaria.

Figure 11   *Radio success stories*

2.6 Reform and revolution

The Mexican revolution in 1910 brought sweeping changes to the countryside. Large estates were broken up and given to poor people.

In 1976, nearly 3 million people were working on land they owned. This is not always the case in Third World countries.

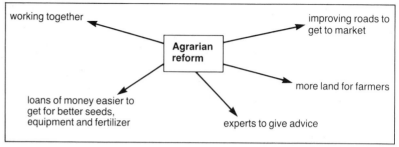

Figure 1 *Agrarian reform*

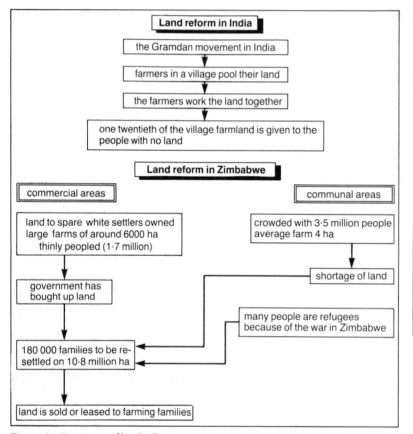

Figure 2 *Two types of land reform*

'Someday all our countryside will have electricity and be filled with towns where there is running water, electricity, gas stoves; and the children won't have to walk two kilometres to go to school in the morning…no longer will clothes be washed in the river and people have to carry water for long distances; the candle and lantern will be no more; and the life of women will be much better.'

Fidel Castro

Figure 3 *Cuba before the revolution*

Change in the countryside

Farmer's lives are affected by who owns the land, how much land there is, how crops are sold and the amount of government help. These things will have to change if people's living conditions are to improve. Such changes are called **agrarian reform** (Figure 1).

One type of agrarian reform is to take unused land or break up large estates owned by rich landowners, and divide it among poor farmers. This is called **land reform** (Figure 2).

Revolution in Cuba

In 1959, a communist revolution in Cuba threw out the landowners and American companies who owned much of the land. They had grown wealthy on cattle ranches and rice and sugar estates. Most Cubans remained poor (Figure 3).

Changing hands

Since the revolution, the government has taken over the land. The large estates and ranches are now **state farms**. Over 50% of the land is worked in **co-operatives**. People share the work and the land. Workers now live in new rural towns with health, education and housing facilities.

There are still over 100 000 small private farmers, but they own only 12% of the land. As they retire or die, the government gets the land. It is not inherited by their families.

The Cuban government has made mistakes and still faces severe problems. But life has changed for the better in the countryside (Figures 4, 5 and 6).

Exercises

(Heading: *Reform and revolution*)

1 What is agrarian reform?

2 Choose one of the land reform projects in Figure 2.
   a Describe how the land changes hands.
   b How will the poorest people benefit?

3 List the ways in which life has changed in Cuba (Figures 3 to 6).

Figure 4   *Farmsteads were isolated with no electricity or water*

Figure 5   *Many people live in new towns; they have electricity, schools, health clinics and TV*

Figure 6   *The government provides equipment from tyres to cane harvesters*

Change in Etah

The Etah district is one of the poorest parts of India. Two million people live in farming villages. Only 22% are literate and their farms are very small. Crop and milk yields are well below the average for India (Figure 7).

Tackling all the problems

Hindustan Lever Ltd is a subsidiary of the Anglo-Dutch company Unilever. It is one of the largest private companies in India. It set up a milk products factory in Etah but found that farmers could not supply enough milk. They needed help to improve their farming. This could not be done without improving all aspects of their daily lives. As a result, problems in farming, health, education and local crafts are being tackled together. They are so closely connected that progress in one cannot succeed without the others. Such a plan is called **integrated rural development** (Figure 8).

Slow to change

After 6 years, the Etah plan had affected 50 villages and 100 000 people. This may seem a small number, but change often has to be slow if it is to be successful. People have to be convinced that change is worthwhile (Figure 8).

Exercises

(Heading: *The Etah project*)

4   In what ways is the Etah project an integrated project? Give examples.

5   Copy down these problems.
     Opposite each, write down the solutions which have worked at Etah:
        low milk and crop yields
        poor land
        lack of seeds and fertilizers
        lack of money

6   What was done to make local farmers try the new ways?

crop yields in quintals (100 kg) per ha		
	Etah	average for India
paddy rice	4·6	12+
wheat	13·0	15+
sugar cane	3000	500+

milk yield per year − very low

500 litres per buffalo

Figure 7   *Low yields at Etah*

the problems
1 poverty
2 lack of money; moneylenders charged too much
3 5% land waterlogged and salty
4 help needed in better farming methods; poor yields and poor quality cattle
5 lack of reliable supplies of seeds and fertilizers

working hand in hand
extension workers talked to farmers
found out their problems and needs
understood local farming conditions
lived in the village

seeing is believing
farms selected for demonstration plots
milk factory has demonstration farm
visits made to outside 'model' farms

standing on your own feet
help given is not free
farmers pay their way
in future they will run things

help to get seeds, fertilizers, loans
help for village craft industries
loans for shallow tubewells for irrigation

training given in farming and equipment maintenance

poorer farmers get loans for better quality buffaloes richer farmers inject their buffaloes with semen of pedigree Jersey or Friesian bulls

cross-bred cattle more expensive to feed higher milk yields

Figure 8 *Problems and changes at Etah*

Working together

Along the coast of south India, many people earn their living by fishing. It is a risky life. Fish only come at certain times of the year. Catches of sardines, prawns and whitebait are small and unreliable (Figure 9).

Families survive by borrowing money from moneylenders for their food and equipment. But moneylenders are also the merchants who buy the fish. They pay as little as possible, so the people stay poor and in debt.

Fishing co-operatives

Fishing families can be helped if they join a fishing co-operative. Several fishing co-operatives have been set up in south India. Families have benefited in a number of ways (Figures 10 and 11). They have more power to improve their lives when they co-operate.

Exercises

(Heading: *Working together*)

7　Make a labelled sketch to show the advantages of the new fishing boat in Figure 11b.

8　How does being in a co-operative keep people out of debt and get them higher prices?

Figure 9　*A small catch of whitebait*

Figure 10　*The traditional fishing boat*

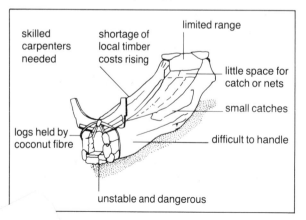

skilled carpenters needed

shortage of local timber costs rising

limited range

little space for catch or nets

small catches

difficult to handle

logs held by coconut fibre

unstable and dangerous

Traditional boat

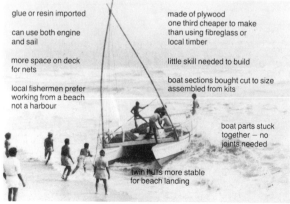

glue or resin imported

can use both engine and sail

more space on deck for nets

local fishermen prefer working from a beach not a harbour

made of plywood one third cheaper to make than using fibreglass or local timber

little skill needed to build

boat sections bought cut to size assembled from kits

boat parts stuck together – no joints needed

twin hulls more stable for beach landing

b　*The sandskipper*

Unit 3 Worlds of change

3.1 Introduction

Electricity is slowly reaching villages in India. Bikes and radios are becoming common, but the motorbike is still a status symbol.

Slow progress

Life changes slowly in developing countries. In villages, skilled men and women still work at home in **cottage industries** such as basket-making and weaving (Figure 1). Traditional work like this is usually the only alternative to farming.

Yet simple improvements can bring big changes to people's lives, such as the new stove shown in Figure 2. With new skills, such as welding, villagers will be able to make and repair their equipment (Figure 3).

Figure 1   *Basket makers in Guyana*

Figure 2   *An improved stove*

Figure 3   *Welding skills in Burkina Faso*

Exercises

(Heading: *Slow progress*)

1   a   What is meant by cottage industry? Give an example.

    b   List three other crafts which you might find in a Third World village.

2   a   Make a sketch of Figure 2. Label on it:
       walls keep heat in...
       higher cooking surface...
       waste cane for fuel...
       fire inside.

    b   Why is this stove an improvement over the open fire which is commonly used for cooking?

3   What evidence of modern ways can you see in Figure 3?

3.2 The right technology

'Give a man a fish and you feed him for a day.
Give him a net and you feed him for life.'

(Chinese saying)

Figure 1   A well pump

Made to measure

Modern machinery, such as a tractor, is not always suitable for use in developing countries. Tractors are expensive to buy. They need proper servicing, use expensive fuel, and sometimes break down. Garages and skilled mechanics are needed when there are problems.

Poor farmers cannot cope with these difficulties. Simple equipment called **appropriate technology** is often more suited to their needs (Figures 1 and 2).

Figure 2   *Thai women pulling a cartload of rice*

Practical and possible

Appropriate technology can be made with local labour and materials. This means that spares are easy to get and repairs are not a problem.

Appropriate technology can solve some very basic problems. For instance, better drying and storage facilities cut down food losses caused by damp and pests (Figure 3). Up to 30% of grain harvests are lost through poor storage.

An improved grain silo

Frame is a woven basket of sticks. Basket plastered with mud, ash and cow dung to make insect-proof. On 4-metre stilts with tins or thorns on the legs to make rat-proof. Thatch of poles and grass with a clay or wooden lid where grain is poured in. Spout made of tin to let out grain. Silo held together by twine and nails.

Takes a person 8 days to make. Lasts 6 to 7 years.

Figure 3   *A grain silo*

Exercises

(Heading: *The right technology*)

1   a   Make a simple sketch of Figure 3.
    b   Make a list of items needed to make the silo.
    c   Tick those you think could be made from local materials.
    d   Why is the silo so suitable for a poor farmer?

2   Study Figure 1 of a well pump. Imagine you were trying to convince some villagers to use one. What main points would you make. (You could draw a poster to illustrate your answer.)

Testing time

Even simple improvements have to be fully tested to see that they work well. Farmers have to be certain that there are no problems.

In 1978, the Sierra Leone 'work-oxen project' was set up (Figure 4). Most farmers in Sierra Leone do not have oxen. The project may be able to show the advantages of using oxen and persuade the farmers to buy them.

Work-oxen project

A detailed study was made of how oxen worked on the land. The same jobs were done by oxen and by hand. They were timed and costed. Usually, oxen were cheaper and quicker, and some jobs like weeding were done better.

The project also studied how to teach farmers to use the oxen. Oxen have to be trained to pull equipment and take commands.

Figure 4   *a   0·5 tonne carried 4 km*

b   equipment like ploughs and seeders can be hooked up to a tool bar

c   oxen trained in 3 weeks; walking with a log is first step before pulling a plough; farmers become ox-handlers in 4 to 6 weeks

d   neck yoke tied to horns make oxen easier to control

f   tool bars can be fitted with various pieces of equipment:
this one has a plough

Maize cultivation activity	time (hours per ha)		cost £ per ha	
	by hand	by oxen and hand	by hand	by oxen and hand
planting	112	22	114	123
weeding (twice)	686	30	686	176
	79	79	79	79
fertilizing	105	105	105	105
harvesting	982	236	984	483

g   oxen save farmers time, money and energy, they are
economical

e   oxen of good size (4 years old) can work 4-5 hours a day for
5 days a week; main job is ploughing from May to June

Exercise

(Heading: *A Report on the Sierra Leone project*)
3   a   Copy the report form Figure 5. Fill in
your own conclusions about the project
using Figure 4.
  b   Do you think oxen would be a big
help to farmers in Sierra Leone?
Give reasons for your answer.

	your comments
Ease of training oxen	
Ease of training men	
Costs compared with a tractor	
Costs compared doing the work by hand	
The use of different types of equipment with the oxen	
Recommendations	

Figure 5   *Report form*

3.3 High fliers

India is a space-age nation. Its first satellite was launched in May 1981. The four stage rocket had 90% of its major parts made in India. This may seem surprising for a country better known for its poverty.

Carts to cosmonauts

The bullock cart is a common sight in India (Figure 1). It gives the impression of a country which has not entered the 20th century.

Yet India also has **high technology** industries such as electronics (Figure 2). India's scientists design advanced modern equipment. Engineers are building nuclear power stations. An Indian cosmonaut has been in space with Russian cosmonauts.

Figure 1 *Bullock cart*

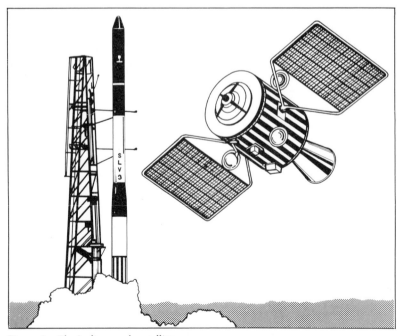

Figure 2 *The Indian apple satellite*

Exercises

(Heading: *Carts to cosmonauts*)

1 a What are high-technology industries?
   b Give an example.

2 In what ways is life in India a mixture of the old and new?

Buying knowledge

Developed countries have the technological knowledge which poor countries lack. Some of this knowledge can be bought. India pays British Aerospace for a licence to make Jaguar aircraft (Figure 3). British Aerospace provides the know-how and some parts so that the plane can be assembled in India.

This knowledge can be used to develop India's own aircraft industry. Eventually India will become **self-reliant** in many products which have to be imported at the moment.

Figure 3   *Making Jaguar aircraft*

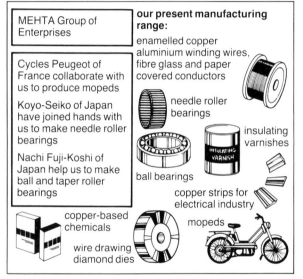

MEHTA Group of Enterprises

Cycles Peugeot of France collaborate with us to produce mopeds

Koyo-Seiko of Japan have joined hands with us to make needle roller bearings

Nachi Fuji-Koshi of Japan help us to make ball and taper roller bearings

our present manufacturing range:
enamelled copper aluminium winding wires, fibre glass and paper covered conductors

needle roller bearings

insulating varnishes

ball bearings

copper strips for electrical industry

copper-based chemicals

mopeds

wire drawing diamond dies

Figure 4   *MEHTA uses foreign know-how*

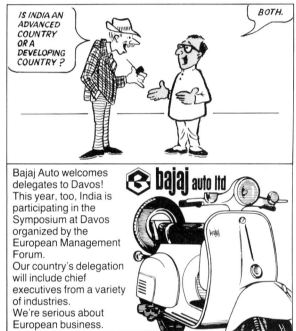

IS INDIA AN ADVANCED COUNTRY OR A DEVELOPING COUNTRY?

BOTH.

Bajaj Auto welcomes delegates to Davos! This year, too, India is participating in the Symposium at Davos organized by the European Management Forum.
Our country's delegation will include chief executives from a variety of industries.
We're serious about European business.

bajaj auto ltd

Figure 5   *The Bajaj scooter*

Exercises

3   How does India get the knowledge to make complicated equipment? Give an example.

4   Figure 6 shows some top industrial companies in India.
   a   Make a list of what the companies make.
   b   Copy these statements correctly (either/or):
       the companies are large/small
       their products cover a wide/small range
       they are similar/different in size to those in developed countries
   c   'India is a poor and backward country.'
       Why is this not completely true (Figures 3 to 6)?

India's Corporate Giants
Figures for 1980–81 in lakhs (100 000s) or rupees

Company	net sales £ MN
Tata Engineering	360
Tata Steel	300
Hind Lever (Chemicals)	260
Delhi Cloth Mills	180
Gwalior Rayon	170
Ashok Leyland	130
Reliance Textile	130
Asso Cement	100

Figure 6   *India's industrial giants*

A nation in a hurry

Figure 6 shows the world's **newly industrializing countries (NICs)**. These countries have rapidly growing manufacturing industries. South Korea for instance is now second to Japan in ship-building (Figure 7).

The South Koreans are also taking a growing share in the world electronics trade. Exports of products such as computers and colour televisions are worth $2 billion.

New industries have helped improve life in South Korea. Though poverty is still a problem, most people have more money, live longer and have better health than 30 years ago (Figure 8).

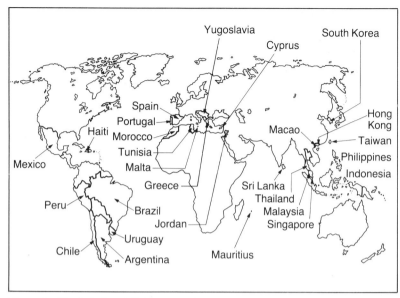

Figure 6 *Newly industrializing countries*

Figure 7 *A South Korean shipyard*

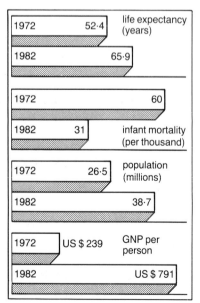

Figure 8 *Life gets better in South Korea*

1972	52·4	life expectancy (years)
1982	65·9	
1972	60	
1982	31	infant mortality (per thousand)
1972	26·5	population (millions)
1982	38·7	
1972	US $ 239	GNP per person
1982	US $ 791	

Exercises

(Heading: *South Korea: a nation in a hurry*)

5 a What is a newly industrializing country?
   b List four examples from Figure 6.

6 Give an example of how South Korea has caught up with other countries.

7 Describe how life has improved in South Korea.

Leading the way

There are many reasons for South Korea's success (Figure 9). Strong governments, the attitude of the people, and US help have all played a part. Third World countries can learn from South Korea's example, but many cannot follow the same path (Figure 10).

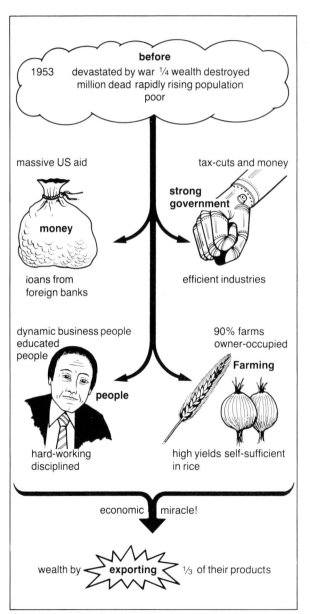

Figure 9 *How South Korea developed*

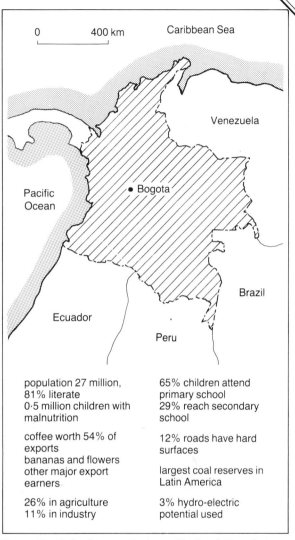

population 27 million, 81% literate 0·5 million children with malnutrition	65% children attend primary school 29% reach secondary school
coffee worth 54% of exports bananas and flowers other major export earners	12% roads have hard surfaces largest coal reserves in Latin America
26% in agriculture 11% in industry	3% hydro-electric potential used

Figure 10 *Colombia – can it change like South Korea?*

Exercises

8 Use Figure 9 to explain why South Korea has some successful industries.
Mention:
    US help...exports...help
    for farming...attitude of the
    people...businessmen...education.

9 Study Figure 10. Why would Colombia find it difficult to industrialize like South Korea?

3.4 New crops

Wheat production went up by four times in India between 1960 and 1980. In Pakistan, wheat and rice production doubled. Growing food has kept ahead of the growth in population because new varieties of those crops were being used.

The green revolution

In the past 20 years, new varieties of crops have been developed. Under the right conditions individual plants of rice and wheat can give a lot more food (Figure 1). The amount per plant is called the **yield**.

These **high-yield varieties (HYVs)** give new hope to poor people. This great change is called the **green revolution**.

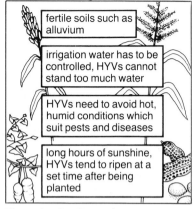

fertile soils such as alluvium

irrigation water has to be controlled, HYVs cannot stand too much water

HYVs need to avoid hot, humid conditions which suit pests and diseases

long hours of sunshine, HYVs tend to ripen at a set time after being planted

Figure 1   *Conditions for HYVs*

Benefits for some

The green revolution has made some regions more prosperous. Richer farmers have often gained the most. They have the money to buy costly fertilizers, tube wells and power pumps. They are more likely to get government loans and be able to transport their increased harvests to markets.

In Sri Lanka, the government encouraged farmers to try the new HYVs of rice. In the first few years, food production rose, but then declined. The government and farmers have learned that HYVs are not an easy short cut to more food (Figure 2).

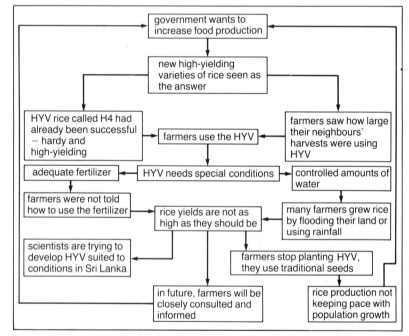

government wants to increase food production

new high-yielding varieties of rice seen as the answer

HYV rice called H4 had already been successful — hardy and high-yielding

farmers use the HYV

farmers saw how large their neighbours' harvests were using HYV

adequate fertilizer

HYV needs special conditions

controlled amounts of water

farmers were not told how to use the fertilizer

rice yields are not as high as they should be

many farmers grew rice by flooding their land or using rainfall

scientists are trying to develop HYV suited to conditions in Sri Lanka

farmers stop planting HYV, they use traditional seeds

in future, farmers will be closely consulted and informed

rice production not keeping pace with population growth

Figure 2   *Seeds in Sri Lanka*

Exercises

(Heading: *Seeds in Sri Lanka*)

1   a   What is the green revolution?
    b   What do the initials HYV mean?

2   Study Figure 2. Give these sentences suitable endings:
    a   Farmers use the HYVs because _____.

    b   The rice did not grow well because _____.

    c   The government was to blame because _____.

    d   Two things being done so that better seeds can be used in the future are _____.

    e   H4 seeds could not be used again because _____.

Crops for cash

It seems odd that poor underfed countries export food to rich overfed countries. The Common Market imports beef from ranches in the drought-stricken Sahel of Africa (Unit 1.4)!

Crops which are grown for sale are called **cash crops**. Sometimes poor farmers have enough food to eat and can sell the surplus. But some cash crops take up land which could be used to feed the local people. In Belize, the spread of sugar has disrupted the way of life of farming families (Figure 3).

The price of sugar rose so farmers planted more. Less food crops were grown. Men took jobs in the sugar fields. Wages gave them money to spend on radios and trucks.

Men did not have the time to cut new fields from the forest. Land was not rested and became more infertile. Yields went down. Maize, rice and wheat had to be imported.

Food crops which were spoiled were fed to pigs and poultry. These gave meat to the families' diet.

Pigs useful to sell when family short of cash. As less food grown so fewer pigs kept.

Women are in charge of food and animals. They now feel less important.

Figure 3 *Sugar leads to bitterness in Belize*

The right crops

Growing one crop in rows in a field makes sense in a developed country. It can be tended easily and harvested by machines. In developing countries, the traditional method of mixing crops in small plots looks untidy. But **intercropping** is a good farming method on many Third World farms (Figure 4).

Exercises

(Heading: *The right crops*)

3 What is a cash crop? Give an example.

4 Read Figure 3, then copy and fill in the diagram in Figure 5. Give it a title.

5 Make a copy of Figure 4, but label the notes in suitable places with arrows.

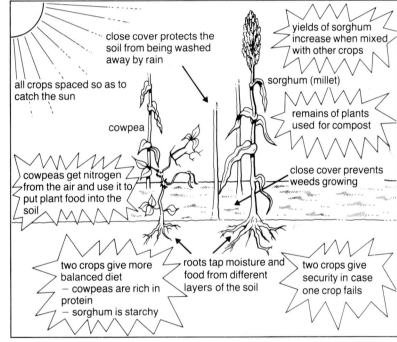

close cover protects the soil from being washed away by rain

all crops spaced so as to catch the sun

cowpea

cowpeas get nitrogen from the air and use it to put plant food into the soil

yields of sorghum increase when mixed with other crops

sorghum (millet)

remains of plants used for compost

close cover prevents weeds growing

two crops give more balanced diet
— cowpeas are rich in protein
— sorghum is starchy

roots tap moisture and food from different layers of the soil

two crops give security in case one crop fails

Figure 4 *Intercropping*

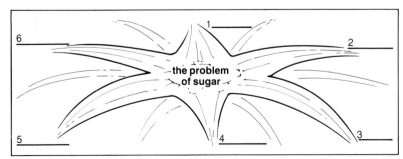

1
6
2
the problem of sugar
5
4
3

Figure 5 *The problems of growing sugar in Belize*

The plantation

Plantations are large farms which use modern farming methods. In many cases they were set up by foreign companies to grow products the company needed at home. Today, however, many plantation products are sold on the open world market. Many plantations grow just one crop (**monoculture**), such as sugar, coffee or cotton.

1 They are modern, commercial (profit-making) businesses.
2 One main crop is grown.
3 Scientific farming methods are used.
4 They are large in area.
5 They have an impact on the surrounding area.
6 A plantation is built as a community.
7 The crop is sold for cash, or is used in manufacturing.
8 The crop is processed in a plantation factory.
9 A large amount of capital is needed to set up the plantation.

Figure 6   *Features of plantations*

The Labuk valley plantation

The Anglo-Dutch company of Unilever developed an oil palm plantation in Malaysia (Figure 7). It began in 1960. The forest was cleared in a remote area 150 km by river and sea from the nearest town. There were no roads or tracks. Local people lived by hunting and subsistence farming. The plantation has changed the area (Figures 8 and 9).

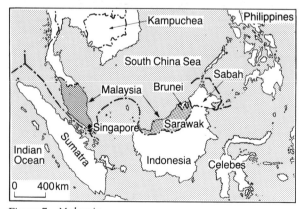

Figure 7   *Malaysia*

Plantation took ten years to develop...employs 2000 people...covers 7000 hectares...split into four parts each with facilities such as housing, shops, health clinic, football field...airfield in the centre...daily services to the nearest town...total cost of £8 million.

Figure 8   *The growth of the Labuk Valley Plantation*

Figure 9 a   *An oil palm plantation in Malaysia*

b new ways of increasing yields; scientists developed the clonal oil palm; root tissue of high-yielding trees is grafted on to young oil palms; yield up 30%

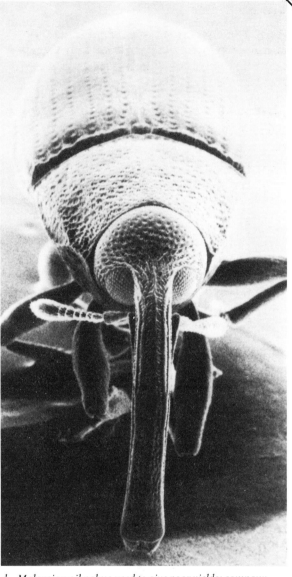

d Malaysian oil palms used to give poor yields; company scientists discovered that a weevil is needed to pollinate (fertilize) the palms

Exercises

(Heading: *An oil palm plantation in Malaysia*)

6   Give evidence to show that it is a typical plantation.

7   Make a simple diagram to show how it has affected the surrounding people.

c   new roads and bridges to improve communications

3.5   Investing in the future

Problems of poverty will not be solved if the conditions which cause these problems are not changed. This takes time and large sums of money. But spending for the future means there is less to spend on the problems of today.

The Majes project

Peru is a large but poor country. Much of the land is either too mountainous or too dry for farming (Figure 1). Growing enough food to feed the rapidly increasing population is a problem.

One way to get more food is to make more land suitable for farming. Dry land can be farmed if it is **irrigated**. This means to bring water to places with too little rain.

The Majes irrigation project is being built to bring water to Peru's desert land (Figure 2). It will cost £750 million and will take 10 years to complete.

This is a lot of money for a poor country such as Peru. The money could be spent now on many smaller projects to help the poor. Instead, it is being spent on a large project which will not bring benefits for many years. Spending money now so as to benefit later is called **investing** in the future.

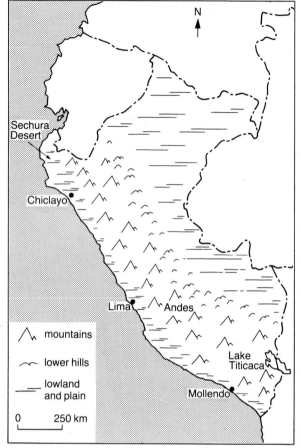

Figure 1   *Peru*

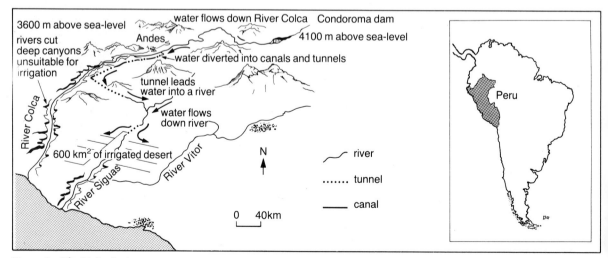

Figure 2   *The Majes Project*

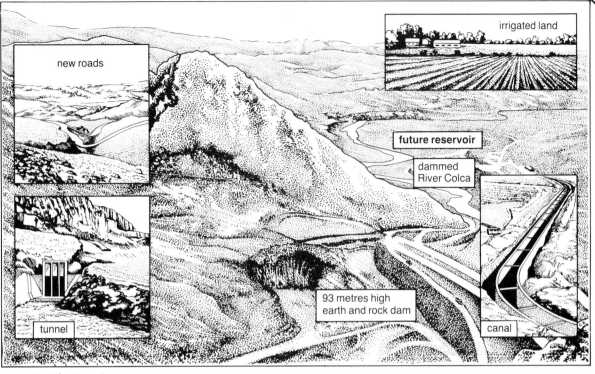

Figure 3   *The Majes Project*

Exercises

(Heading: *The Majes project*)

1   Use Figure 2 to make a simple sketch of the project. Mark on:
dams...canals...tunnels... mountains...heights above sea level ...irrigation areas...Pacific ocean ...the rivers Colca and Siguas ...the town of Arequipa.

2   Copy out the paragraph below and fill in the blanks. Use these words:
desert...power...Andes...Majes power stations...Colca...60 000 ...enormous...tunnels.

Nearly 5000 people worked on the _____ project. The waters of the river _____ high up on the _____ mountains are being diverted through 99 kilometres of _____ and 23 kilometres of canals. These bring the water to the _____ 250 kilometres to the coast. The cost of the project is _____ but the benefits will be great. About _____ of new farmland will be gained. Up to 50 000 jobs will be created in industry. In addition, two hydro-electric _____ _____ will be built. These will bring an 800% increase in Peru's electricity supply. People will have electricity in their homes for the first time and factories can use it for _____.

3   What reasons would you give a poor farmer for not spending the project money to help people now?

4   Make a short description of the project by linking up the sketches in Figure 3.

Factory in the forest

The largest industrial plant in central Africa is being built at Edea in the Cameroons (Figure 4). This is the Cellacum pulp mill (Figure 5). It is jointly owned by the Cameroons Government and foreign companies who have invested money in it.

The plant will use wood from the surrounding tropical forest. Logs will be broken down to a cardboard-like material called pulp. Paper and card are made from the pulp.

There are 350 species in the area and unlike other forestry schemes in tropical areas, all of these species will be used at Cellacum.

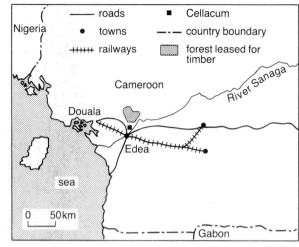

Figure 4   *The Cameroons*

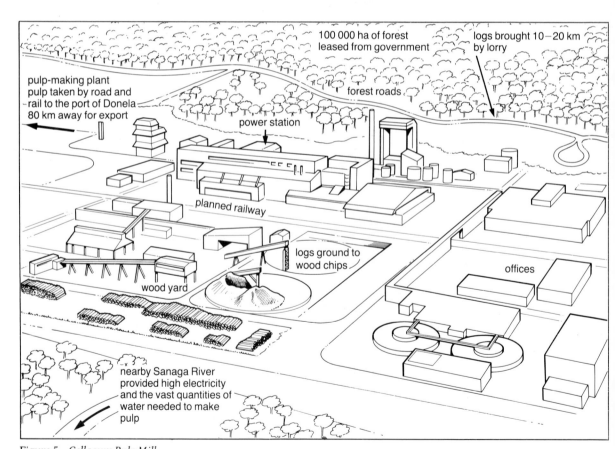

Figure 5   *Cellacum Pulp Mill*

Finding more jobs

The mill is planned to produce 120 000 tonnes of pulp a year for the next 30 years. For its size and cost, few workers will be employed at the factory. Machines do most of the work (Figure 6). A project like this is called **capital intensive**.

But exporting the pulp will bring money into the country. There will also be other industries and new jobs because of the pulp plant (Figure 7).

Figure 6 *Machines in the forest*

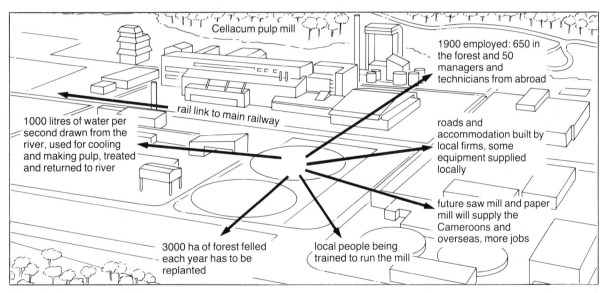

Cellacum pulp mill

1900 employed: 650 in the forest and 50 managers and technicians from abroad

rail link to main railway

1000 litres of water per second drawn from the river, used for cooling and making pulp, treated and returned to river

roads and accommodation built by local firms, some equipment supplied locally

future saw mill and paper mill will supply the Cameroons and overseas, more jobs

3000 ha of forest felled each year has to be replanted

local people being trained to run the mill

Figure 7 *The impact of Cellacum*

Exercises

(Heading: *A factory in the forest*)

5 What does the Cellacum plant do?

6 The location of the plant was carefully chosen. Describe what things were taken into account. Mention:
the forest...river...access ...exporting.

7 a Why is it a capital intensive project?
  b What are the advantages of such a project?

8 Draw a labelled sketch to show the advantages and disadvantages of the Cellacum project. Mention:
labour...use of the natural forest ...worker training...road and rail access...new industries.

A trickle of jobs

In Bangladesh, 42% of the workers are unemployed and 50% are landless. Work employing large numbers of people is needed. Such work is called **labour intensive**.

Traditional handicraft industries such as woodcarving and shoe making are examples of labour intensive jobs (Figure 8). The products are made by hand using local skills and materials (Figure 9).

If products are to sell in local markets or abroad, they have to be well made and carefully chosen. Charity organizations such as Tradecraft and Oxfam give money to help start these local industries. They can also help with advice on which products and designs will sell best (Figure 10).

Exercises

(Heading: *A trickle of jobs*)

9   a   Why is the job in Figure 9 labour intensive?
      b   Give two reasons why these are suitable jobs for a developing country.

10   a   Copy out this sentence: Village craftsmen make things which are essential to the lives of the people.
      b   List three other craftworkers (apart from shoemakers and woodcarvers) you would expect to find in Third World villages.

Figure 8   *Mexican shoemaker*

Figure 9   *Cane furniture as a craft industry*

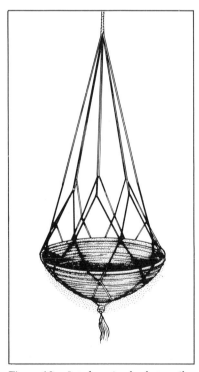

Figure 10   *Jute hanging basket or sika*

Finding customers

It is often hard to get products to where they can be sold (Figure 11). In Bangladesh, rickshaws, buses and trains are all used.

Exporting is especially difficult. Workers do not have contacts in the government and in shipping companies. Organizations like the Jute Works help with this (Figure 12). They also arrange for the goods to be sold in shops in the rich countries.

Small beginnings

Small rural job creation projects cannot solve the unemployment problem. Only a dozen people may be involved in a project (Figure 13).

The most needy are helped first. These are often women such as widows who are the family wage earners. They work at home, in small groups, or as part of a co-operative.

New products

Products often go out of fashion or are no longer needed. This means that workers have to switch to new products which will sell. Job creation programmes in the Third World aim to provide jobs that will last.

Exercises

(Heading: *Finding customers*)

11 Use Figure 11 to explain why it is difficult to get products to market.

12 Describe how the Jute Works helps its members get orders. Mention: designs...transport...forms... exporting...shops.

13 How will job creation projects help increase the number of workers (Figure 13)?

The village is only 64 km from the capital of Bangladesh. But it takes a day to get there! This is the journey:

45 minutes by road in a minibus
60 minutes along a dirt track
30 minutes to cross a river by ferry
10 minutes walk while the ferry is lifted over a dam
another 2 hours up river by ferry
a 2½ km walk to the village

Baskets are made in the village co-operative. Getting them to a market is very difficult.

Figure 11 *Getting to a remote village*

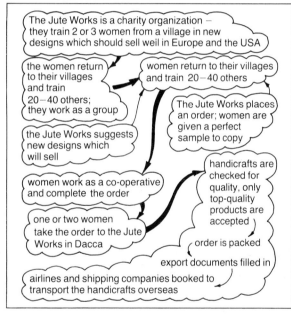

Figure 12 *How the Jute Works helps*

The workshop in Hyderabad employs about 20 men. They are being trained to make shoes and sandals. They are paid 50p a day (10 rupees) and work nine hours with a one hour break. When trained, the men return to their homes and teach others. The men used to be rickshaw pullers earning up to 80p for a 12-hour day. They worked in all weathers and had no proper breaks. They found it hard to feed their families or keep themselves strong for such tough work.

Figure 13 *A new job for the rickshaw pullers*

3.6 Change in China

China aims to become one of the world's leading industrialized countries in the next 20 years.

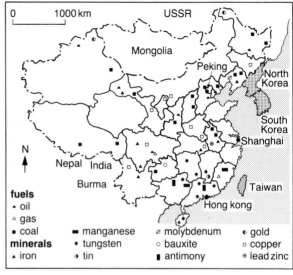

Figure 1   China's resources

Riches for the future

Developing countries are poor in many ways, but rich in others. China has abundant supplies of materials people need. These are called **resources**. China has world class deposits of several resources such as tin, bauxite and iron. Oil and coal production are rising rapidly and there are plans to open 29 new coal mines (Figure 1). The resources are being used to expand China's manufacturing industries (Figure 2).

Giant steps forward

China's people are better fed than 30 years ago, when famine and starvation were common. Now China grows most of its own food, though only 11% of the land is suitable for growing crops (Figure 3).

light industry	millions	heavy industry	millions
yarns (tonnes)	3·36	rolled steel (tonnes)	29·02
cotton cloth (metres bn)	15·35	other steel (tonnes)	37·16
chemical fabrics (metres bn)	4·56	cement (tonnes)	95·20
silk fabrics (metres)	914·00	sulphuric acid (tonnes)	8·17
sugar (tonnes)	3·38	chemical fertilizer (tonnes)	12·78
beer (tonnes)	1·17	rubber tyres	8·64
salt (tonnes)	16·38	mining equipment ('000 tonnes)	158·00
bicycles	24·20		
sewing machines	12·86		
wrist watches	33·01		
television sets	6·21		
household washing machines	2·53		

Figure 2   Industrial production in China

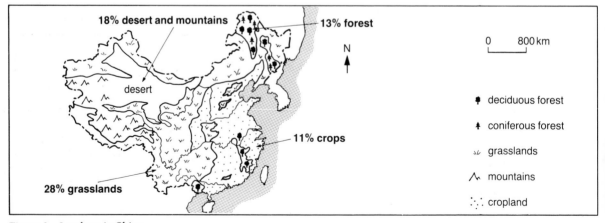

Figure 3   Land use in China

Rivers under control

Rivers are the lifeblood of China. Nearly half of all freight is carried on waterways, and 47% of cropland is irrigated. But for thousands of years, the rivers brought devastating floods and were little use in time of drought.

The Huang He was called China's sorrow because of the death and damage it caused. Now the river is controlled with dams and reservoirs (Figure 4). It has not seriously flooded for 25 years. But as in other developing countries, some disasters still happen (Figure 5).

Figure 4   *Sanmenxia Dam*

reservoir to cover the land and homes of 600 000 people

90 m high

summer flood discharge reduced from 37 000 m^3/s to 8000 m^3/s

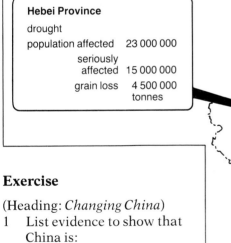

Hebei Province

drought

population affected	23 000 000
seriously affected	15 000 000
grain loss	4 500 000 tonnes

Peking

Shanghai

Wuhan

Hong Kong

Hubei Province

flood

population affected	20 200 000
seriously affected	6 200 000
houses damaged or destroyed	510 000
grain loss	2 590 000 tonnes

Exercise

(Heading: *Changing China*)

1 List evidence to show that China is:

　　rich in resources

　　developing manufacturing industry

　　difficult to farm in many areas

　　putting large rivers to good use

　　not free of disasters

Figure 5   *Disasters still happen*

People at work

In China, people do work that is done by machines in developed countries. Millions of people are used to carry out big projects (Figure 6). In 1958 for example, 1 million people raised the embankments along the Huang He river by 1 m over a distance of 600 km.

Twenty million bikes

The bicycle is the cheapest and most convenient form of transport in China. Over 20 million bikes are produced each year compared with 200 000 vehicles, 500 trains and 40 000 tractors. The bike is China's answer to the car. It carries families, baskets of vegetables, pigs and even television sets (Figure 7).

Figure 6   *Digging a new channel in the Huang He river*

Figure 7   *A land of bikes*

Steaming ahead

About 44% of all freight is transported by rail. This is much larger than in developed countries. Most engines are still powered by steam. This makes sense in a country rich in coal where electrification of the railway track is too expensive. Steam trains are the appropriate technology for China to use (Figure 8).

Figure 8   *The Datong steam locomotive factory; locos cost £100 000 each to make; factory turns out over one per week*

Exercise

2   Explain why it is more suitable in China to:
     use people instead of machines
     use bikes rather than vehicles
     use steam trains

Unit 4   Signs of strain

4.1   Introduction

'History is bunk'   **(Henry Ford 1919)**

Model growth

There is nothing new about poverty. Even in today's rich developed countries, there was appalling poverty less than 100 years ago. By building up industry and by trading all over the world, a much higher standard of living has been achieved. Models of development have been produced to show how this has happened (Figure 1).

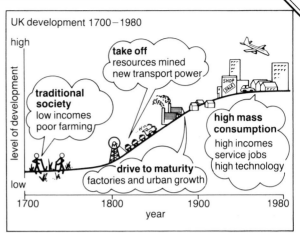

Figure 1   *A model of development*

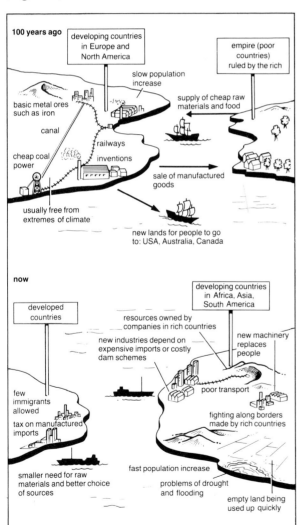

Figure 2   *Development in a changing world*

Lessons from history

With this in mind, it is easy to think that all the poor countries need to do is follow the same path to development.

But history does not have all the answers. Poor countries today are in a very different situation from countries such as the UK 100 years ago (Figure 2).

Getting worse

In many ways, the problems facing developing countries are getting worse rather than better. The gap between rich and poor is increasing year by year.

Exercises

(Heading: *Towards development*)

1   a Make a copy of Figure 1.
    b What is the problem in using this model to help poor countries today become richer?

2   a In what ways were today's developed countries able to use the poor countries to make themselves rich?
    b Use the bottom part of Figure 2 to describe some of the problems which poor countries have to face.

4.2 The numbers game

Figure 1 *A happy event*

The birth of a child is usually a happy family event. There are about 200 000 happy events every day all over the world. This is adding up to a local, national and international nightmare.

Too many, or too few?

Not every government thinks that **population control** is needed in their country (Figure 2). They think that too few people hold back progress. There may not be enough people to work on farms, in factories or build new roads. With too few people, it is not worthwhile making mass-produced goods for sale.

Greater numbers may be needed for defence against a nearby country with a much larger population.

The problems of increase

Fortunately, 81% of the people in developing countries live where the government wants to slow down the increase (Figure 3). They see how too many people cause problems in housing, employment, health and education.

	%		%
Bulgaria	0·4	Iraq	3·4
France	0·5	Israel	1·7
East Germany	0·0	Kampuchea	1·9
Greece	0·7	Laos	2·4
Luxembourg	0·0	Mongolia	2·9
Chad	2·0	Argentina	1·6
Gabon	1·2	Bolivia	2·7
Guinea	2·5	Chile	1·5
Ivory Coast	2·9	Uruguay	0·8
Libya	3·5		

Figure 2 *Countries which want to increase population (% increase each year)*

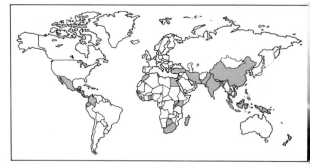

Figure 3 *Countries which want to reduce population growth (shaded on the map)*

Exercises

(Heading: *Population control*)

1 What do Figures 2 and 3 tell you about each of these things:
   a the number of countries in the world who do not want to control population numbers compared with those that do;
   b the rates in increase in countries which do not want population control;
   c the location of countries that do not want to control population increase;
   d the location of countries which want to limit their population?

2 How might a shortage of people affect each of these things:
   a starting an industry such as making cars (*hint*: skilled workers and customers);
   b exploiting new resources in a remote part of the country (*hint*: labour);
   c building new roads to link places where people are very spread out (*hint*: cost);
   d defending the country (*hint*: manpower)?

The Indian census

The results of the 1981 population **census** in India were unexpected. The census showed there were 12 million more people than the government knew about. There has been a population control plan since 1952, but the growth is still much too fast (Figure 4).

total (1972)	550 700 000
total (1982)	713 800 000
increase %	2
birth rate	35
death rate	15
time to double	35 years

Figure 4   *The Indian census*

Control by law

The Indian government cannot make a law to limit the number of babies born. India is a **democratic** country where people vote for their politicians. The people would not agree to such a tough law.

Instead, the law has been used in other ways. In 1978, the legal age for marriage was raised from 15 to 18 for women, and from 18 to 21 for men. Later marriages should mean fewer children. Abortion has been legal since 1971, though only in cases approved by a doctor.

Information

There is a nationwide campaign persuading people to have fewer children. This is done by posters, TV, radio and cinema advertising, and even by puppet shows. People are told that two children are enough (Figure 5).

Making it work

Many methods of **birth control** are being tried. The pill, sterilization and devices such as the coil are used to prevent pregnancy. But there are still too few medical staff and health centres for these methods to reach everybody.

Figure 5   *Population control: this poster announces a 'small family' competition—only those who have had a vasectomy are eligible*

Exercises

(Heading: *The Indian census*)

3   Why do you think it is so hard to keep an accurate check on the number of people in a developing country such as India? Think about these things:
   the size of the country
   the number of people
   the speed of change
   communications
   recording and administration

4   a   Explain why a law telling people how many children they can have would not work in India.

b   In what ways is the law in India used to control population increase? Explain how each law would work.

c   Describe any other ideas you can think of which might make people have fewer children, for example, tax laws.

5   a   List the methods being used to persuade people in India to have fewer children. Say which you think would reach the most people and which would be most effective.

b   Why is it so hard to bring birth control methods to most of the Indian people?

area	9 561 000 km^2	
population	759 600 000	1972
population	1 000 000 000	1982
increase %	1·4	
birth rate	22 per 1000	
death rate	7 per 1000	
time to double	48 years	

Figure 6   *China's population statistics*

The Chinese billion

China is the first country in the world to have more than 1 billion people. This is almost one in every five people on earth (Figure 6).

In China, 60% of the people work in farming. There is less than 0·1 ha for each person. There are 25 cities with over 1 million people.

China is not a rich country so it is little wonder population control is taken very seriously.

Aiming for zero

The aim is to have the same number of births as deaths. This would mean a **stable** population total with **zero growth**.

Already the growth rate is down to 1·2%. This is the lowest figure for any large developing country. The percentage increase is less than 1% in 10 out of China's 15 provinces.

Good health for all

China is a **communist** country. The government can spend money on whatever it thinks is most important. A good health service is one of their main aims. Figure 7 shows how the system works.

There are 1·8 million **barefoot** or Red guard health workers in China. They are trained to give first aid and teach health care. They also help with contraceptive advice and equipment. Some even perform abortions.

Better health care means people do not need to have so many children in case some die.

Exercises

(Heading: *Population in China*)

6   Find a map of China in your atlas which shows population density.
  a Draw a simple sketch map to show where the people live. Mark in the main cities.
  b Write a paragraph to say how population density in China compares with other parts of Asia, Africa and South America.

7   a What is meant by the term zero growth?
  b What figures tell you that population control is being successful in China?

8   a How does the fact that China is a communist country help bring better health to the people?
  b How has the Chinese health service got over the problem of shortage of doctors?
  c Explain how a good health service and population control are linked.

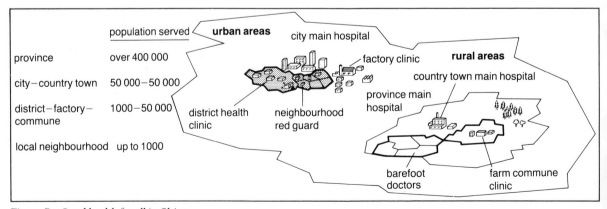

Figure 7   *Good health for all in China*

Figure 8 *Being glorious and having one child*

Figure 9 *The Glory certificate*

Helping each other

Chinese farms are run as co-operatives with each family helping the other. This means there is no need to have so many children to do the work. Old people are looked after by the **community** so they do not have to depend on their children.

Education reaches both old and young. The need for population control is explained so that everyone understands what is expected of them (Figure 8).

The one child family

Parents with one child get a certificate which gives them special benefits (Figure 9). There is better housing, free health and education, and extra money. The child is given top priority when finding a job.

If there is a second child, these benefits are lost. There is a special tax on people who have more than two children.

Every area is given a population **quota** (number) by the government. Local officials try to see that the quota is kept.

Breaking traditions

Many ancient customs have had to be changed. Marriage is not allowed before the age of 20 for women and 22 for men. In some districts, the figures are 25 and 28 years.

Women now have equal rights to men with equal pay and job opportunities. There is child care in creches and nurseries. This encourages women to stay in work and have fewer children.

The Chinese way

Some people say that the Chinese rules about children are too strict. The government interferes too much in people's rights to have children. Too much **freedom** is lost.

The answer from China is that people who have no food, bad health and no future, have no freedom anyway. At least these problems are being solved in China.

Exercises

(Heading: *Population under control*)

9 Explain how each of the following helps limit the number of children in China:
   co-operative farming
   education
   child benefits
   child quotas
   marriage laws
   equal opportunities

10 What do you think about the laws on population control in China. Are they too strict, or is there no other choice?

4.3 The need for land

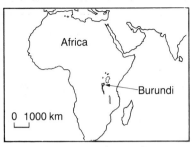

Figure 1 *Location of Burundi*

It is said that everyone in the world could stand on the Isle of Wight. This may be true, but they could not grow enough food to live on in this amount of space.

Small, steep and rural

Burundi is one of the smallest countries in Africa. It is a little smaller than Wales, and much of it is just as steep. Unlike Wales, 98% of the people live in rural areas with 84% of the jobs in farming.

Figure 2 *Farming landscape in Burundi*

Nothing wasted

The photograph shows a typical scene in Burundi (Figure 1). Farming looks impossible on such steep slopes. Yet vegetables, beans, bananas and coffee are grown in areas like this. No space is wasted.

So far so good

The traditional method of farming is to **cut and burn**. First the trees are cut down. Then the land is cleared by burning.

After about 4 years farming, the field is left to become overgrown again and a new area is cleared. The land should be allowed to rest for at least 30 years before being used again.

Exercises

(Heading: *Farming in Burundi*)

1   Copy and complete the paragraph below using the following words. You need an atlas. Zaire...Africa...Victoria...equator...high... south...Rwanda...Tanzania...metres. Burundi is a country in _____. It lies just ____ of latitude 0, the _____. Most of the country is _____ land with the highest peak at 4507 _____. The countries of _____, ____ and _____ are its nearest neighbours. Lake _____ is about 150 km to the east.

2   Make a labelled sketch of Figure 2. Include the % of people farming, the types of slopes, methods of farming and what is grown.

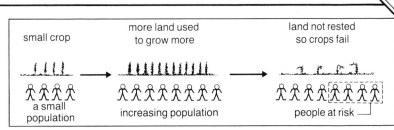

total	4 000 000
increase %	2·2
birth rate	45
death rate	23
density in 1972	129 per km²
density in 1982	143 per km²

Figure 3 *Population in Burundi*

Figure 4 *Food production and population*

People and land

As well as being one of the smallest countries in Africa, Burundi is also one of the most **densely populated** (Figure 3). There are 235 people to every km² of farmland. Each family has on average less than 1 hectare to grow the food they need. On a farm no larger than three football pitches, barely enough can be grown to keep a family healthy.

Gifts from God

Like in other African countries, the population total is rising quickly. At 2·2% increase each year, 4 million people in 1982 will become 8 million in 30 years time. One survey in Burundi showed that women want an average of 8 children in their family. Children are seen as a gift from God as well as being useful helpers on the farm. But more children means even less land and less food for each person.

Soil is destroyed

Already there are signs that land is being **overused**. Land is being used after only 2 or 3 years rest. More food has to be grown for the increasing population.

Soil that has not rested does not become fertile again, so new crops do not grow properly (Figure 4). In Figure 5, there is a deep **gully** where rain has washed the soil away. This happens when there are no trees to protect the soil during heavy rainstorms. Gulleys become deeper and wider as long as the land is left unprotected. All this means even less land, and a growing demand for more food.

Figure 5 *Soil erosion*

Exercises

(Heading:*Ruining the land*)

3 a Copy the information in Figure 3.
  b Write a paragraph using this information to explain why it is becoming more difficult to feed the people in Burundi.

4 a How has population increase changed the way land is farmed in Burundi?
  b Draw 4 sketches to show how a gulley is formed on a steep hill slope.
    i Cutting down trees
    ii Farming the land
    iii Heavy rainstorms
    iv Soil washed away
  c What are the effects on crops of not leaving the land to rest?
  d Use the information in Figure 4 to explain how food production has kept pace with population increase so far. Say what might happen in the future.

A treeless world

The story of Burundi is not unusual. Trees in many developing countries are also being felled. This is called **deforestation**. Little is being replanted (Figure 6).

Since 1950, half the world's **tropical rain forest** areas have been felled. Two thirds will be gone by the year 2000. Trees in the drier **savanna** areas are also being cut down.

Demand for wood

Clearance for farmland is the main reason for deforestation. Clearance for firewood is another reason. One third of the world's population rely on firewood for cooking and heating. These people live mostly in the developing countries (Figure 7).

There is a demand for tropical woods such as teak and balsa from rich countries. Selling wood is one way that poor countries can earn money from rich countries.

From bad to worse

Unprotected soil is either washed away by rain, or in drier areas, blown away by the wind. Savanna land becomes desert when too many people try to grow too much and graze too many animals. This is called **desertification** (Figure 8).

Figure 6   *Forest destruction*

Figure 7   *Collecting firewood is women's burden*

Exercises

(Heading: *World deforestation*)

5   a   Write down two facts which show how quickly deforestation is happening.
   b   Explain the link between deforestation and each of these things:
      new farmland...fuel...exports.
   c   Why do you think that so little tree planting is taking place in developing countries? Think about:
      the need for farmland
      the cost of planting trees
      the time, labour and skills needed
      government laws in rural areas

6   a   What type of land is usually affected by desertification?
   b   What does Figure 8 show?
   c   Where is desertification happening? Name some places.
   d   Explain what causes desertification.

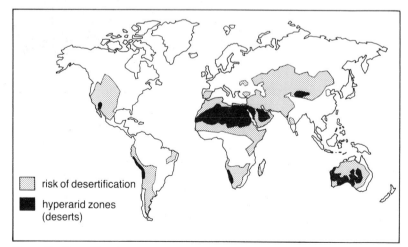

risk of desertification

hyperarid zones (deserts)

Figure 8   *Desertification*

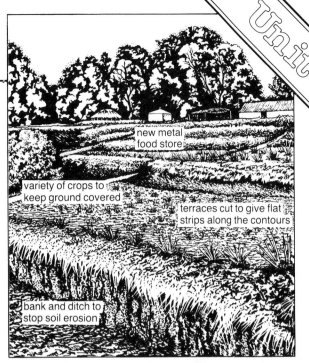

Figure 9 *Slope management in Burundi*

Thinking about today

Often, peasant farmers do not understand the problems caused by overusing the land. Even when they do, there may be nothing they can do about it. Land cannot be left unused while the family go hungry.

Slope management

Figure 9 shows some ideas which help stop **soil erosion** in a tropical area such as Burundi. Flat terraces with ditches and banks stop water rushing down the slopes. Adding compost helps keep the soil fertile. Some crops give nitrogen to the soil through their roots (Figure 10). Tree planting protects soil until it is used.

Against the desert

In savanna areas, tree planting, new wells, and more care over the number of animals can stop the land becoming desert (Figure 11). Ideas like these need money for materials, education for the people and skilled people to show how they work. In any new plan, local people have to be involved and their wishes taken into account.

Exercises

(Heading: *Making improvements*)

7 a Use Figure 9 to draw a cross-section down a slope. Add labels from Figure 9.
   b Write a paragraph to say how these new ideas can help the Burundi farmers grow more and make better use of their land.

8 a Use Figure 11 to draw a map of how a savanna area could be planned so that land is not overused.
   b Describe any problems in making the plan work.

Figure 10 *Learning new farm methods*

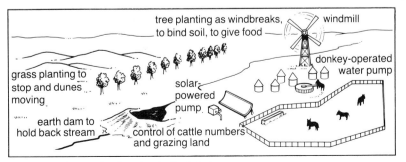

Figure 11 *Improving farming in the Sahel*

4.4 Work and migration

'Men go on saving labour till thousands are without work, thrown on the open streets to die of starvation'

(Gandhi 1924)

Vanishing jobs

In both rich and poor countries, jobs are vanishing faster than new ones are being created. In the poor countries, population increase makes jobs even harder to find.

Off the land

In country areas, many farms are now so small that it is hard to grow enough to make a living. In countries where land is divided among all the sons, farms are now much too small. This has happened because so many children now survive. Land cannot be divided any further. Some end up with nothing.

Owners of small farms cannot afford to hire labourers. On large farms, profits are spent on machines such as tractors (Figure 1). One tractor can do the work of at least 5 workers.

A job that pays

The answer for many is to move to the cities. Moving from one place to another is called **migration**. But city jobs are just as hard to find.

Jobs in poor countries that pay a living wage are even harder to find (Figure 2). In São Paulo, a factory worker earns only enough to survive in a rough shack. Differences in the cost of living in each country do not explain the different wage levels.

Figure 1 *Work for tractors or people?*

Take it or leave it

When so many people are looking for work, employers can pay low wages and still get enough workers. Few poor countries have social security payments for the unemployed. If wage rates rise, machines can be bought to replace the workers.

People in rich countries benefit from low wages in poor countries. Low wage rates mean cheap exports to developed countries.

Exercises

(Heading: *Loss of jobs*)

1   How does the population increase affect the number of jobs in a country. Why is this a special problem in developing countries?

2   Explain why each of these things causes a loss of jobs:
    a subdividing land,
    b mechanization (on farms and in factories).

3   a Give some examples of different wage rates for the ame job in different countries.
    b Is it fair that wages are so different?
    c How do people in rich countries benefit from low wages in poor countries?

net incomes of occupational groups			(where London = 100)		
	building worker	fitter/ turner	textile worker	teacher	secretary
Jakarta **Indonesia**	13	15	6	15	44
Sao Paulo **Brazil**	35	71	34	34	169
Milan **Italy**	68	67	104	104	80
Tokyo **Japan**	141	172	167	167	213

Figure 2 *The rate for the job*

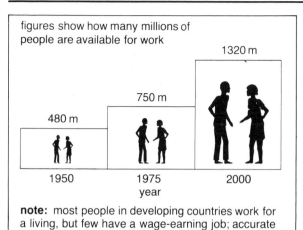

figures show how many millions of people are available for work

1320 m

750 m

480 m

1950   1975   2000
year

note: most people in developing countries work for a living, but few have a wage-earning job; accurate unemployment figures do not exist

Figure 3   *World unemployment*

Figure 4   *Informal jobs in a city*

Unemployed and underemployed

There are said to be 450 million people in poor countries either unemployed or **underemployed** (Figure 3). Many of the underemployed work in **casual jobs** where they are hired by the day. Often there is no work to do.

Service jobs

In the cities, many make a living by providing a service. Examples include cleaning shoes or working as servants for rich families (Figure 4). Some recycle scrap materials they find around the city. Up to 70% of the people in some cities earn a living in these kinds of jobs. These jobs are part of the **informal** sector.

A question of politics

Getting better wages and more jobs is often a matter of politics. Trade Unions are illegal in many countries so workers cannot organize themselves. Higher wages for some would mean lower profits for the landowners and factory owners. These are often the same people that run the country. Wealth can only be shared when the government really wants this to happen.

Exercises

(Heading: *The world's unemployed*)

4   a   What do each of these terms mean?
            unemployed...underemployed
            ...casual or informal jobs.
     b   What does Figure 3 tell you about
            unemployment now and in the future?

5   a   Make a list of about 10 service jobs you
            think you could do in a city.
     b   Is there anything you think you could
            make out of scrap materials? Where
            would you go to get the materials?
     c   What are the problems in making a
            living in the informal sector? Think
            about:
                the amount of money that can be
                  earned
                how regular the work is
                competition from others
                what about the future?
                is it legal?

6   Explain why it is so hard for people in
       many poor countries to get better working
       conditions and wages. Make sure to
       mention laws, ownership and politicians.

country of origin	number
Mexico	52 100
Philippines	41 300
South Korea	29 200
China/Taiwan	24 300
Vietnam	22 500
Jamaica	19 700
India	19 700
Dominica	17 500
Cuba	15 600
UK	13 400

Figure 5   *Migration to the USA*   Figure 6   *Illegal immigrants*

Migrants to the USA

About 300 000 people migrate to the USA every year (Figure 5). Many of these people come from developing countries. The largest numbers come from the Caribbean islands and from Mexico.

The USA is the nearest rich country to these places. It is where there is the best chance of a better life.

Over the border

Most Mexicans go to California (51%) and Texas (21%). Some come with green cards which allow them to live and work in the USA as permanent **aliens**. An unknown number cross the border illegally (Figure 6). Illegal entry is often organized by a coyote, the name given to a smuggler. Mexicans also cross the border to work every day, then return home at night.

Migrant workers

Migrant workers do all kinds of work (Figure 7). Some are skilled and professional people looking for better opportunities. Most take the low paid jobs that people in the USA do not want to do.

Workers are needed on farms and in the food processing factories. This work is seasonal and the workers return to Mexico when the job is done.

Migrants also work in the cities, mostly in low paid jobs such as dishwashers or cleaners. There are jobs in small workshops making clothes. In these jobs, skilled hands are needed more than machines.

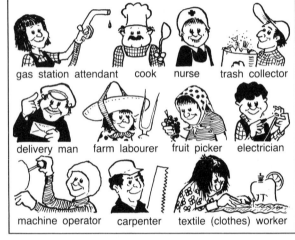

gas station attendant   cook   nurse   trash collector

delivery man   farm labourer   fruit picker   electrician

machine operator   carpenter   textile (clothes) worker

Figure 7   *Typical migrant jobs in the USA*

Exercises

(Heading: *Mexican migrant workers*)

7 a Migrants usually move from poor areas to the nearest rich area. How does Figure 5 show this to be true?

   b What are the three ways by which Mexican workers enter the USA? Use these words in your answer:
      green card...alien...illegal... coyote...daily.

   c Draw a sketch map from your atlas to show these places:
      Mexico...the USA...Texas... California...Rio Grande river.

8 a Why do skilled and professional Mexicans sometimes want to leave Mexico?

   b Give some examples of the kinds of job Mexican migrants do which are:
      low paid...seasonal...skilful.

Migrant problems

There are about 1 million people who were born in Mexico now living in the USA. Most are proud of their past and do not want to become full citizens of the USA. This would mean taking out **naturalization** papers. Spanish is still spoken at home and Mexican customs are important to them. Not everyone welcomes Mexican migrants and their families and sometimes there are problems (Figure 8).

Flight from Nigeria

Some migrations have ended in disaster. In 1982, the Nigerian government decided that illegal immigrants had to leave quickly. A few years earlier, oil had made the Nigerian economy grow quickly. Nobody minded the arrival of illegal workers because they helped in the growth.

Get out!

But the price of oil fell and jobs in Nigeria became scarce. Migrants were blamed when Nigerian workers could not find a job. The result was that about 1 million Ghanaians were told to leave. Thousands more from other West African countries were also ordered out.

   They crowded on to lorries and into ships (Figure 9). There were deaths from disease and exhaustion. Migrants became the victims of problems they did not cause.

Figure 8   *Some problems of migration*

Figure 9   *Ghanaians go home*

Exercises

(Heading: *Migrant problems*)

9   a   Who is likely to be most badly affected by migrant workers?
   b   Who benefits most from migrant labour?
   c   The USA and other countries put a limit on the number of migrants allowed in each year. Give reasons why this is done.

10   Write a short newspaper article to tell the story of what happened in Nigeria. Invent a suitable headline. Include the main facts of what happened. Write your own comment on what you think about the Nigerian government's decision.

4.5 Cities in crisis

Cities in developing countries are growing so fast that figures are soon out of date. Some have doubled in size in as little as ten years, and will soon double again. Life in these cities is already a fight for survival for most of the people. The chances of improvement seem small.

1980	population in millions	estimate for AD 2000	population in millions
New York	16	Mexico City	31·0
Tokyo	15	Sao Paulo	25·8
London	11	Tokyo	24·2
Shanghai	10	New York	22·8
Mexico City	8	Shanghai	22·7
Los Angeles	8	Beijing	19·9
Buenos Aires	8	Rio de Janeiro	19·0
Paris	8	Bombay	17·1
São Paulo	8	Calcutta	16·7
Osaka	7	Jakarta	16·6

(cities shaded are in developing countries)

Figure 1   *World top ten cities*

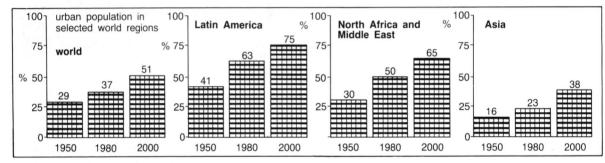

Figure 2   *Urbanization of the continents*

Millions and millions

In 1950, there were only 6 cities in the world with over 5 million people. By 1980 there were 25 (Figure 1). By the year 2000, there will be at least 60, of which 45 will be in the developing countries.

The largest cities will have more than 10 million, with some even over 20 million.

The urban world

The percentage of people living in cities is increasing (Figure 2). This change from rural to urban areas is called **urbanization**.

Most of the increase is because people from rural areas are moving to the cities. But as the total world population is also increasing, there will still be many more millions living in rural as well as in urban areas.

Exercises

(Heading: *World urban growth*)

1 a Draw a simple world outline. Mark in the world's top ten cities in 1980 (Figure 1).

   b On a second outline, mark in the top ten cities for the year 2000.

   c What was the average size of the top ten cities in 1980? How does this compare with the average size for the top ten in AD 2000? How do UK cities compare?

2 Copy and complete these sentences with a suitable ending:

   a The increasing percentage of people in cities is called _____.

   b Most of the world's largest cities in AD 2000 will be in _____.

   c Cities are growing so fast because people from rural areas _____.

   d The growth of cities will keep increasing as long as the world population _____.

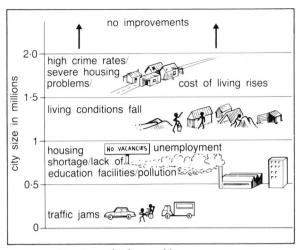

Figure 3   *City size and urban problems*

Figure 4   *Klong Toey slum district*

Breaking point

As cities grow, the problems become greater (Figure 3). In poor countries, there is not enough money or skills to solve these problems. It is the poorest people who are most affected.

Bangkok grows

The growth of Bangkok in Thailand is typical for a developing country. From 1 million in 1950, there are now 6 million people, with 12 million forecast for the year 2000.

Life in a swamp

There are 400 slum districts in Bangkok with at least 1 million people living in them. About 400 000 people live in the Klong Toey district. One visitor described Klong Toey as an area of 'tin and wooden shacks over a swamp of black mud, untreated sewage and rotting rubbish' (Figure 4). Drinking water comes from the rivers and public taps. Both are badly polluted from factories and sewers.

Sinking fast

Water is pumped from rocks below the city. Because of this, the city is sinking by up to 12 cm per year in some places. As most of Bangkok is only 1 m above sea level, monsoon rains and high tides make flooding a common problem.

Social problems

There is high unemployment and problems of crime, drugs, disease and prostitution. About 50 000 children under 15 years old are thought to work illegally in factories. Some work for up to 18 hours a day. In a city this size, laws are hard to enforce.

Exercises

(Heading: *Big city problems*)

3   Find Bangkok in an atlas. Draw a sketch map to show where it is.

4   Use the information in Figure 3 to describe the problems that occur as a city becomes larger.

5   a   Draw and label, or describe Figure 4. Include these things:
        where it is
        how many people live in the district
        the streets
        the type and condition of the houses
    b   Describe the following problems:
        water supply...flooding...employment ...health.

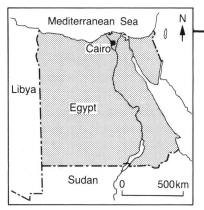

Figure 5   *Location of Egypt*

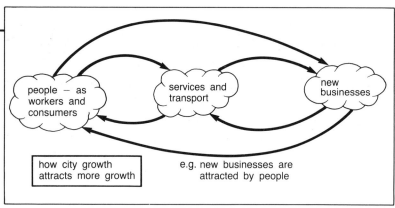

Figure 6   *The multiplier effect*

The heart of Egypt

Cairo is the capital and largest city in Egypt. With 12 million people, it is also the largest city in Africa. It is six times larger than Alexandria, the next largest Egyptian city. A main city like this is called a **primate city** (Figure 5).

Cairo has grown at an average of 4·1% each year in recent years. This is much faster than Egypt's 3·1% increase. One in every four Egyptians now lives in Cairo.

Growth feeds growth

Cairo began in AD 969. It has been a major trading, manufacturing, cultural and political centre for most of its history.

Today, it is the obvious place for new industries and offices to go (Figure 6). It is the main communications centre in the country. The large population means that there is a large labour force and many people to sell goods to (Figure 7).

It is the best place to go for other reasons. One in every three doctors in Egypt works there. The universities and colleges have 65% of all Egypt's students.

Figure 7   *Modern city life in Cairo*

Figure 8   *A farm in Egypt*

A matter of choice

Conditions in rural areas also help explain the growth (Figure 8). There are 32 000 villages in Egypt. In most, the water supply is bad and health and education facilities are few. In many of the smaller towns, conditions are little better. As a result, people move to Cairo from all the settled parts of Egypt (Figure 9).

A shortage of land

Eight out of every ten farms in Egypt are smaller than 1·25 hectares. This is barely enough to support a family. As the population increases, the amount of land for each family gets smaller. In Islamic law, land is divided equally among the sons, so farms have become very small.

The growing cities cover about 8500 hectares of extra land every year. This is usually on fertile Nile valley farmland.

Moving to the cities is the only choice left for country people who want a better future.

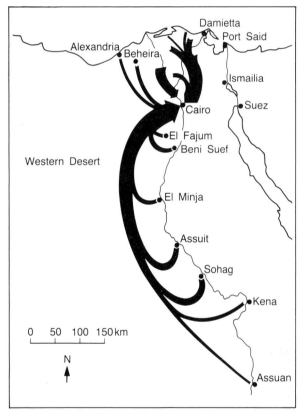

Figure 9   *Movement of people to Cairo*

Exercises

(Heading: *The growth of Cairo*)

6   What is a primate city and what makes Cairo an example of one?

7   Explain each of the following:
   a   Cairo is the obvious place for new businesses to go,
   b   there are opportunities for a better life in Cairo.

8   Use Figures 8 and 9 to explain why so many people have been moving to Cairo from other parts of Egypt. Mention:
   where people are moving from
   living conditions in rural areas
   problems of land shortage

Ready to burst

The average living density in Cairo is 26 000 people per km^2. The most crowded districts have up to 110 000 per km^2. In 1979, about 2 million new or improved houses were needed. Although 85 000 were built in that year and more have been built since then, the housing problem remains enormous.

Building up

One answer to the housing problem is to build tall blocks of flats (Figures 10 and 11). But these are expensive, and take too long to build. The people most in need of housing cannot afford to either buy or rent them.

Exercise

(Heading: *Building flats*)
9   a   What facts tell you that there is a serious housing problem in Cairo?
    b   Describe the type of building shown in Figures 10 and 11.
    c   Why will this type of building not solve the housing problem?

Low cost answers

One idea being tried in Egypt is to build **low cost housing**. Sometimes, as in Figure 12, only the land is provided, the people build the houses themselves. Sometimes a concrete base with water and sewerage is provided. This is called a **site and service** scheme.

    Most of the poorest people live in the **shanty town** districts on the edge of Cairo. They are not able to afford even the cheapest houses. For these people, improving the houses they have built for themselves might be the best answer. This is called **slum upgrading**.

    The house shown in Figure 13 is a core unit. At first, only two rooms are built. Other rooms can be added on later if the family can save enough money.

Figure 10   *Cairo's new skyline*

Figure 11   *New flats on the edge of Cairo*

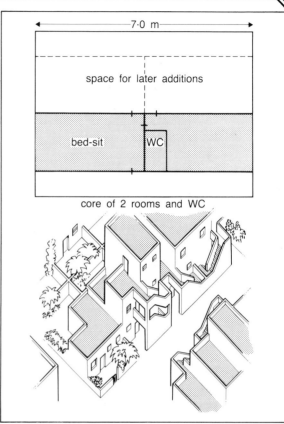

Figure 13   *Core unit housing*

Figure 12   *Low cost housing in Egypt*

Exercises

(Heading: *Low cost answers*)

10 a Describe or sketch Figure 12.
   b What are the advantages of this type of housing scheme?

11 a What is a shanty town and where are they usually built?
   b Why are many types of low cost housing not much help to people who live in shanty towns?
   c What is meant by slum upgrading. Why is this kind of scheme needed in developing countries such as Egypt?

12 Draw a sketch of the house in Figure 13. Label in what is provided, and how the house can be extended.

Into the desert

Egyptian planners are trying to stop even more people moving to Cairo. They are also trying to get people to move out of the city. To do this, five **new towns** are being built in the desert about 50 km away from Cairo (Figure 14). About 3 million people are to live in these new towns.

Plans for the cities

There are growth plans for other cities in Egypt such as Alexandria, Suez and Port Said. New industries are helped with government money. **Free trade zones** have been opened where industry is free from many normal taxes. Growth in these cities may divert growth from Cairo.

The root of the problem

There are rural improvement schemes and plans to build 4000 new villages with modern facilities (Figure 15). Factories for processing farm produce are being built. These ideas should help keep more people in the rural areas.

Above all, population increase must be controlled. The present rate of 1 million extra people every year has to be reduced if plans for both urban and rural areas are to work.

Exercises

(Heading: *Planning for the country*)

13 a Find a map of Eygpt in an atlas. Draw a large outline of the country and mark in these places:
　　　Cairo...Alexandria...Suez...Port Said
　　　...Aswan...River Nile...Libyan desert.
　　b Add notes in boxes to show how Egyptian planners are trying to control the growth of Cairo. For example,
　　　new villages with modern facilities,
　　　new towns in the desert,
　　　free trade zones in ports,
　　　factories in rural areas.

Figure 14　*New housing in one of Egypt's new towns*

Figure 15　*A village in Egypt*

　c Explain why controlling population growth is so important in any plans to improve living standards for people in Egypt.

4.6 Unequal shares

In Guatemala, 2·2% of the people own 70% of the land. In Mexico, one in ten own 40% of the country's wealth. The poorest 20% have only 3% of the wealth. In the Third World, it is often true that most wealth is owned by a few.

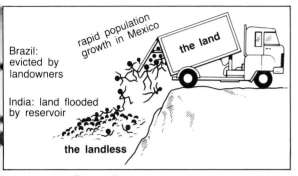

Figure 1 *Landless on the increase*

Land for a few

Land in developing countries is usually owned by a few landowners. They may have so much land that some of it is left unused. Poor farmers cultivate it illegally and are called **squatters**. Others rent the land as **tenants**. Some tenants called **share croppers** give part of their harvest as payment to the landowner.

Millions of people have no land to farm at all. These are the **landless**. One in four of the workers in India and 10 million people in Brazil are landless labourers. The army of the landless is growing (Figure 1).

The rich in poor countries

The rich are well educated and often copy the western lifestyle. Some are millionaires. Many have made their money in business or are politicians. Some have even grown rich through corruption. They have a powerful say in the running of the country. Most live in the attractive suburbs in cities.

The rich are a favoured group of people called an **élite**. Their way of life often cuts them off from the poor. Governments try to satisfy their needs, and this may work against the poor (Figure 2).

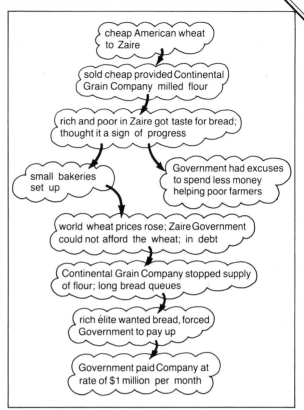

Figure 2 *Grain comes to Zaire*

Exercises

(Heading: *The rich in poor countries*)

1 Copy out and match these pairs correctly:

Renting land from a landowner	squatters
More wealthy and owning land	sharecroppers
Farming land illegally	landowners
Pay part of the harvest to the landowner..................	tenants

2 How do people in poor countries get rich?

3 What happened to the following in Figure 2?
   the Continental Grain Company
   poor farmers
   the Zaire government
   rich people

Figure 3 *Women are mothers and workers*

Figure 4 *Driving machines*

women take charge of selling their produce: 35 kilo loads as well as a baby to a market up to 25 km away

men will only do jobs which pay a wage; if there are none they will spend their time chatting and drinking palm wine

women grow food; they dig and sow and weed up to 8 hours a day; fields may be 10 km away and they may carry 35 kilo of firewood home for an evening's cooking

men work the machines

men cut down trees to clear new land

men build the huts

women carry water from up to 4 km away; the pots may hold 10 litres

girls marry and have children between the ages of 16 and 18; they carry on working

Figure 5 *Women at work in Zaire*

Women at work

In many developing countries, women work from 5 a.m. till 7 p.m., a working day of 16 hours. They do two jobs. At home they look after the children, cook and clean. In the fields, they look after the crops (Figure 3). Like the men, women 'age' quickly. A 30 year old can look 50.

The benefits of being male

Men too work long hours, but they often do the more valued jobs. They have wages as in mining or working on estates. They plough the land or grow crops for cash. Many new ideas and methods help men rather than women (Figure 4).

Exercises

(Heading: *Women at work*)

4　a　Study Figure 5, then make a list or series of sketches to show the jobs done by women in Zaire. Give your work a title.
　　b　Describe why you think that women suffer from exhaustion and age quickly.
　　c　List what men do.
　　d　Why do they choose these types of work?

5　Look at Figure 4. What things do women miss out on?

6　'Women should have equal rights to men.' Look at Figures 4 and 5. Explain why you agree or disagree with this statement.

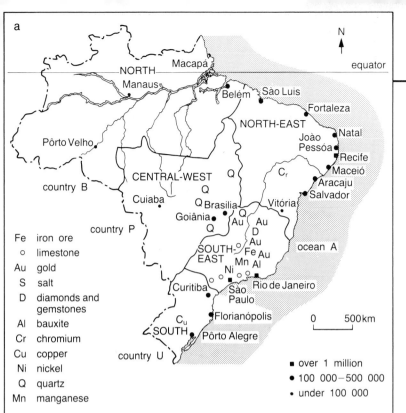

a

Fe — iron ore
o — limestone
Au — gold
S — salt
D — diamonds and gemstones
Al — bauxite
Cr — chromium
Cu — copper
Ni — nickel
Q — quartz
Mn — manganese

■ over 1 million
● 100 000 – 500 000
• under 100 000

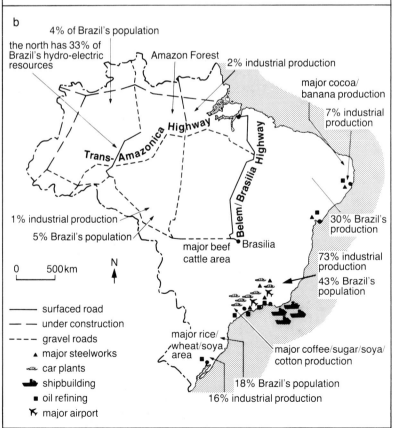

b

— surfaced road
-- under construction
--- gravel roads
▲ major steelworks
🚗 car plants
🚢 shipbuilding
■ oil refining
✈ major airport

Land of contrasts

In both area and population, Brazil is one of the largest countries in the world. It is also one of the richest in **resources**. It is a major world exporter of coffee, iron ore and sugar. Farms, industries and roads have grown up because of these resources.

These developments are unevenly spread. A large part of Brazil's industry, population and wealth is in the south east. This region also has more schools and hospitals than elsewhere.

Parts of the country which are particularly well developed are called **core regions**. In contrast, Brazil's north and interior are much less developed and have fewer people (Figure 6b).

Exercise

(Heading: *Land of contrasts*)

7 a Make a trace of Figure 6a.
   b Name the places marked.
   c Place your trace over Figure 6b and mark on the extra information.
   d Describe how your map shows that the south east is the core region of Brazil. Mention:
        transport facilities,
        mineral and food resources,
        population.

Figure 6 *a Brazil: people and regions*

b Brazil: resources and development

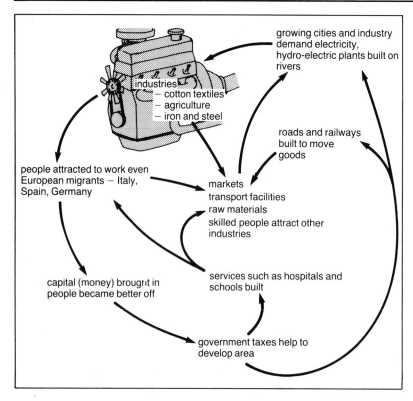

growing cities and industry demand electricity, hydro-electric plants built on rivers

industries
 – cotton textiles
 – agriculture
 – iron and steel

roads and railways built to move goods

people attracted to work even European migrants – Italy, Spain, Germany

markets
transport facilities
raw materials
skilled people attract other industries

capital (money) brought in people became better off

services such as hospitals and schools built

government taxes help to develop area

Figure 7   How growth in the south east snowballed

new housing and industrial centre near Salvador
dams along the San Francisco river provide irrigation and hydro-electricity

building the new capital city of Brasilia
giant hydro-electric plant at Itaipu

building of trans-Amazon highways opening up of manganese and iron ore mines
forest cleared – pasture now for 5 million cattle

Jari River project (rice farming, pulp mill and cattle ranch) has attracted 25 000 people, free education and medical care for families

Figure 8   Projects for the fringe areas

Moving ahead

The south east has a number of advantages over other regions. The climate and soil are suitable for successful farming and there are many mineral resources. The first roads and railways were built there. All these advantages gave the south east a head start. Once the south east started developing, it had a snowballing effect (Figure 7). It now has some built in advantages because of this early lead.

Helping the fringe

The Brazilian government is trying to develop other regions. New highways are making Amazonia more accessible. Irrigation schemes in the north east are helping to combat drought. But the south east still attracts most new jobs and investment. The north and central areas remain on the fringe of developments. (See Figure 8.)

Exercises

8   Describe how the diagram in Figure 7 works.

9   a   Make a list of the projects designed to help the more backward parts of Brazil (Figure 8).
     b   Underline in different colours those concerned with:
           making the region more accessible
           producing more food
           exploiting mineral resources
     c   If these projects are successful, what effect will they have on:
           people leaving the countryside for the cities;
           the size of population in the fringe areas;
           the quality of life of people in these fringe regions?

Unit 5 Choices and conflict

5.1 Introduction

There are many ways in which a country can try to become more developed. But difficult choices have to be made. It is easy to make mistakes which can do more harm than good.

The Mahaweli project

In Sri Lanka, four dams and their reservoirs will provide irrigation water and hydro-electric power for thousands of people (Figure 1).

This scheme will triple Sri Lanka's supply of electricity and 1 million people will receive year-long irrigation on their farms. Yet like all large-scale projects, the Mahaweli scheme has drawbacks as well as benefits (Figure 2).

The reservoirs will flood land. The Victoria Dam's reservoir will cover the homes and land of 40 000 people. They will be resettled on new land.

Another problem is the cost of the project. The original cost was $137 million, but this has risen to over $2000 million! It is very difficult for a poor country to raise this amount of money.

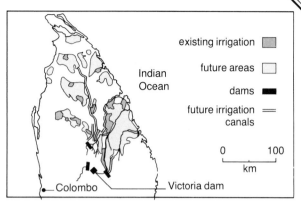

Figure 1 *Dams to provide power*

Exercises

(Heading: *The Mahaweli project*)

1 Study Figure 2.
   a Find evidence to show that these statements are true:
      widespread changes will take place
      some people may not benefit
      fuelwood will be lost
   b Explain why the following will happen:
      new industry attracted
      more food grown
      the government will get more money

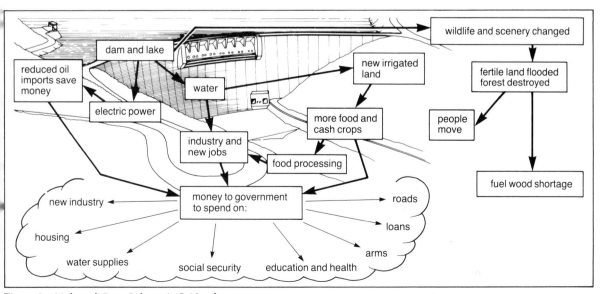

Figure 2 *Mahaweli Dam Scheme in Sri Lanka*

5.2 Aid

Aid from rich countries to poor countries sounds like a good idea. Like all good ideas, aid has to be thought out very carefully both by the giver and by the receiver.

The givers

Figure 1 shows where **aid** money comes from. Most comes from the governments of rich western countries. Organizations such as Oxfam and Save the Children Fund do valuable work, but give only a small percentage of the total.

The receivers

There are few poor countries which do not receive aid from someone. Figure 2 shows the top ten countries that received aid in 1981. But some of the poorest such as Bangladesh are not on the list. Israel is much richer but is second on the list!

The amount of aid each country receives depends on its historic and political links with rich countries. Countries without rich friends do very badly.

	aid per head US $	index of wealth GNP per head in US $
Surinam	320.0	2360
Israel	211.8	3730
Papua New Guinea	101.0	620
Jamaica	77.7	1190
Somalia	60.0	not available
Senegal	42.8	360
Zambia	37.6	510
Burkina Faso	33.1	165
Niger	31.9	240
Sri Lanka	30.1	210

Figure 2   *The top ten receivers (measured in aid per person)*

How aid is given

About 70% of aid goes directly from one government to another. This is called **bilateral aid**. Most of the rest goes through international agencies such as the United Nations. This is called **multilateral aid**.

	percentage
OECD countries (Organization for Economic Co-operation and Development: mainly rich western countries)	68.5
OPEC (Organization of Petroleum Exporting Countries: mainly oil rich Middle East countries)	20.6
Communist countries (mainly the USSR and other East European countries)	5.5
Non-government organizations (voluntary groups such as OXFAM)	5.4

Figure 1   *The aid donors*

What aid is spent on

Many aid schemes have already been mentioned in this book. The dam scheme in Peru (Unit 3.5), farm improvements in Burundi (Unit 4.2) and new housing in Cairo (Unit 4.5) are three examples. Figure 3 shows how aid money is spent.

projects	building new roads, dams, hospitals, water supply
programmes	mainly money to help pay for imports
technical	skilled people sent to give advice and to train local people
food	food sent to make up shortages and to keep prices low
emergency	food, medical aid and supplies to disaster areas

Figure 3   *How aid money is spent*

Exercises

(Heading: *International aid*)

1   Explain each of these terms:
    western countries
    bilateral aid
    multilateral aid
    international agencies

2   What does the information in Figures 1, 2 and 3 tell you about each of the following:
    a   which countries give the most aid;
    b   which countries get most aid, and why;
    c   different kinds of aid?

Boomerang aid

About half the aid given is **tied**. This means it has to be spent on goods and services from the country giving the aid (Figure 4). In this way, much of the money comes back to help the economy of the rich country. Poor countries do get some benefits, but because the aid is tied, they do not always get the kind of help they really want.

Take care

Many aid projects seem like a good idea, but can cause problems to those they are trying to help. Figure 5 shows how this can happen.

Trickle down or fall behind?

It used to be thought that if the whole economy of a country improved, jobs and better wages would **trickle down** to reach the poorest. This does not always happen.

Those who own land get the profits from new crops and irrigation. More money is spent in urban rather than in the rural areas where most people with the greatest problems live. Aid can mean the rich get richer and the poor fall even further behind.

Figure 4   *Boomerang aid*

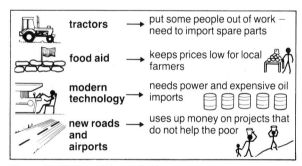

Figure 5   *Some problems of aid*

Exercises

(Heading: *Problems with aid*)

3   a   Write a paragraph to explain the title Boomerang aid.
    b   Why do you think that rich countries prefer to give aid that is tied? Here are some clues:
        public opinion and votes
        jobs
        exports (future spare parts)
    c   Describe how tied aid might not be the most suitable kind of help for a poor country. Think about machinery or an expensive engineering project. Use Figure 5 for some more ideas.

4   How is it that people who are most in need do not always benefit from international aid projects? Mention these things in your answer:
        the cost of large engineering schemes
        fees paid to officials, builders and
          designers
        who owns the land or factories
        the effect of more machinery in factories
          and on farms
        where in the country is most aid money
          spent

Village India

Almost 80% of India's people still live in the countryside, mostly in very poor conditions. In spite of new industries and more wealth, the Indian government still does not have enough money to solve the problems on its own. Aid schemes are going to be needed for many years to come.

Village in a valley

Umaigandhi village is home for 24 families with 106 people (Figure 6). The photograph shows the villager's everyday world, a valley that can only be reached after a five hour walk from the nearest town (Figure 7). The villagers' problems are like those of villagers all over India.

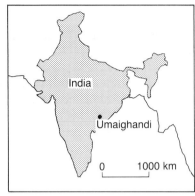

Figure 6   *The location of Umaighandi*

Trapped by poverty

All improvements cost money, whether for new crops, animals or building materials. But the villagers are never able to save any money. Everything is spent on buying their basic needs.

health...water-borne disease common

education...nobody goes to school, too far away

women only speak a local language that is not written down

remote

soil erosion after years of use

lack of regular water, depends on monsoons

no nearby drinking water

customs...people do not drink cow's milk

Figure 7   *Umaighandi Valley*

Exercises

(Heading: *Umaigandhi village study*)

5   Describe the problems caused by each of the following:
   a location of Umaighandi;
   b climate;
   c relief and soils;
   d size of the village population.

6   The villagers grow and make most of what they need. They earn a small amount of money by selling things at their nearest market town.
   a What kind of things do you think they would need to buy at the market?
   b How does this shortage of money stop them making improvements to their way of life?

Aid in action

Aid money has been used to help the people of Umaighandi to get started. A regular water supply was needed to irrigate the fields (Figure 8). This meant building a small dam across a nearby river, then digging a channel to the fields. A new 15 metre deep well in the village saves much time and effort.

Aid money paid for the materials, but an Indian engineer designed the dam and channel (Figure 9). The villagers themselves did the work.

Fruit and goats

Fruit trees are now being grown on the valley slopes. This protects the soil and gives a new supply of food. A rural co-operative centre has given each family three goats (Figure 10). This gives the villagers milk, butter and cheese. When the first kid is born, it is given to the co-operative who then gives it to people in another village.

Figure 8   *The dam site*

Only a start

The villagers are still poor, but a small start has been made to change this. They now know there is help from outside the village if they need it. They also know that their own efforts can solve some of their problems.

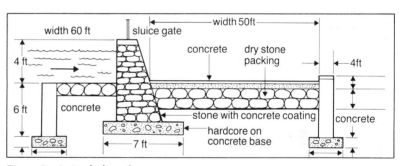

Figure 9   *A simple dam plan*

width 60 ft
sluice gate
width 50ft
concrete
dry stone packing
4 ft
4ft
6 ft
concrete
stone with concrete coating
hardcore on concrete base
7 ft
concrete

Figure 10   *Uses of goats*

Exercises

(Heading: *Aid for improvements*)

7   Use Figures 8 and 9 to write an engineer's report on how the village water problems could be solved. Divide your report into these sections:

water resources
dam construction
water diversion
drinking water supplies

8   How do each of these things help make life better for the villagers:
a   fruit trees;
b   goats?

5.3 Working worldwide

Cadbury Schweppes is a major UK food company. It has branches in 15 countries, both developed and developing. In 1979 the company made profits of £100 million, over half from sales abroad. Standard Oil of New Jersey USA is even larger with more wealth than Indonesia, one of the world's most populated countries.

The big boys

Very large wealthy companies have offices, factories or farms in many countries (Figure 1). They are called **transnationals** or **multinationals**. They employ thousands of people and invest millions of dollars in foreign countries. Transnationals have a variety of business activities. Volkswagen for example, not only make cars but also own cattle ranches.

Decisions made by these companies affect the economy of countries in many parts of the world.

Exercises

(Heading: *The big boys*)

1   a On a world map, plot the countries where Eastman Kodak has businesses (Figure 1).
    b Why is transnational a suitable name for Eastman Kodak?

2   Study Figure 2.
    a In what other ways is Lonhro a typical transnational? Mention:
        has a wide variety of business activities
        large numbers of people employed
        it plays an important role in Africa
        it is important to some countries
    b Name 4 businesses which are found in the UK and overseas.
    c In what ways do you think Lonhro,
        helps developing countries,
        makes some people there uneasy?

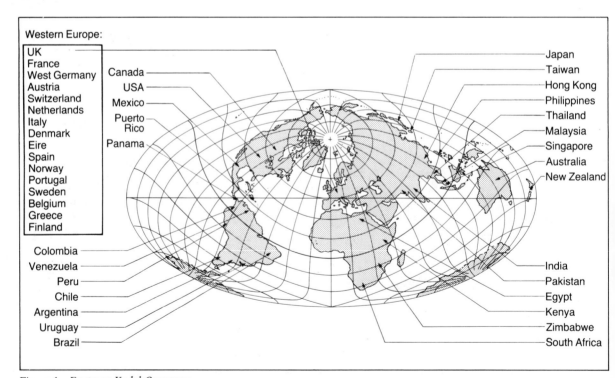

Figure 1 *Eastman Kodak Company*

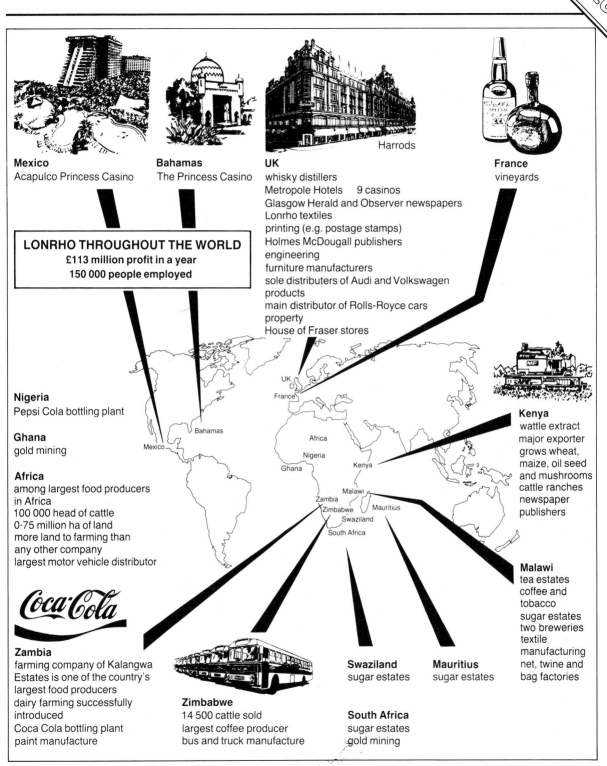

Mexico
Acapulco Princess Casino

Bahamas
The Princess Casino

UK
whisky distillers
Metropole Hotels   9 casinos
Glasgow Herald and Observer newspapers
Lonrho textiles
printing (e.g. postage stamps)
Holmes McDougall publishers
engineering
furniture manufacturers
sole distributers of Audi and Volkswagen products
main distributor of Rolls-Royce cars
property
House of Fraser stores

Harrods

France
vineyards

LONRHO THROUGHOUT THE WORLD
£113 million profit in a year
150 000 people employed

Nigeria
Pepsi Cola bottling plant

Ghana
gold mining

Africa
among largest food producers in Africa
100 000 head of cattle
0·75 million ha of land
more land to farming than any other company
largest motor vehicle distributor

Kenya
wattle extract
major exporter
grows wheat, maize, oil seed and mushrooms
cattle ranches
newspaper publishers

Malawi
tea estates
coffee and tobacco
sugar estates
two breweries
textile manufacturing
net, twine and bag factories

Zambia
farming company of Kalangwa Estates is one of the country's largest food producers
dairy farming successfully introduced
Coca Cola bottling plant
paint manufacture

Zimbabwe
14 500 cattle sold
largest coffee producer
bus and truck manufacture

Swaziland
sugar estates

South Africa
sugar estates
gold mining

Mauritius
sugar estates

Map labels: UK, France, Bahamas, Mexico, Africa, Nigeria, Ghana, Kenya, Zambia, Malawi, Zimbabwe, Mauritius, Swaziland, South Africa

Figure 2   *Lonrho throughout the world*

Uranium at Rössing

The Rio Tinto Zinc Corporation (RTZ) runs the world's largest uranium mine at Rössing in Namibia (Figure 3). The uranium ore is processed, then transported to countries such as the UK for use as nuclear fuel. In 1982, RTZ made a profit of $32 million from the mine.

Better lives at Rössing

The mine has brought many benefits to the people of Namibia. At a time when one in three black Namibians are unemployed, the mine provides work for over 2000 people. Although most managers are white, Namibians can train and gain promotion.

Workers and their families live in housing built by RTZ at Arandis, a new town 12 km from the mine. It has facilities like those found in small towns in rich countries (Figure 4).

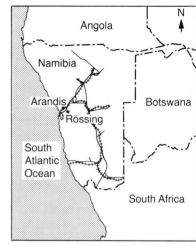

Figure 3 *Namibia*

all workers are covered by accident insurance and receive sick pay

mine workers have a 41-hour week

further education classes in needlework, health, gardening and cookery

a women's cooperative making clothes and leatherwork

scholarships can be won to study in overseas universities

workers at the mine can train for a variety of jobs ranging from office workers to engineers; promising people are sent to university; the aim is for black Namibians to help run the mine

the new town of Arandis has an open-air cinema, shopping centre, swimming pool, library, housing with hot water and electricity, a 40-bed modern hospital, a sports complex and community centre, two primary schools and a secondary school

Figure 4 *a Arandis new town*

Not all agree

Namibia is illegally ruled by South Africa. The United Nations have demanded that Namibia is allowed to become independent. A guerilla movement called SWAPO has grown up and South Africa has to keep 100 000 troops in the country to keep control. The troops have also been used to break up strikes (Figure 6).

The government of South Africa runs the country through a policy called apartheid. As a result, black and white people have to live separate lives. Black Namibians have few opportunities to better themselves. Opponents of South Africa believe that RTZ should leave Namibia and is helping South Africa by staying (Figure 5).

There is also concern at the health and living conditions of the mine workers. Uranium is a radioactive substance and can cause cancer. Many miners are migrant workers. Some have had to leave their families behind and live with other men in crowded dwellings.

Figure 5 *'They couldn't rule themselves anyway'*

job grade	dollars per month		numbers of employees			total
	min.	max.	black	coloured	white	
1	296	317	166	5		171
4	400	460	322	48	10	380
8	712	1068	51	104	59	214
11	1097	1646	0	10	148	158
13	1499	2548	1	1	47	49

Figure 4 *b A pay chart*

Exercises

(Heading: *Conflict over Rössing*)

3 Study Figure 4.
   a Describe the attractions of living in Arandis.
   b What are the advantages of working for RTZ?
   c Do black and white people get equal pay for equal work?

4 List all the things which concern people about the Rössing mine.

5 Study Figures 5 and 6. What are the cartoons saying about life in Namibia?

Figure 6 *Keeping control*

5.4 Trouble from trade

By 1914, France, Germany and Britain controlled over half the world. Europeans went to settle in countries all over the world.

Echoes of the past

The countries controlled by Britain made up the **British Empire** (Figure 1). They were called **colonies**. Colonies like India and Nigeria helped make Britain a prosperous and powerful nation. They provided the British people with cheap food, and British industry with raw materials.

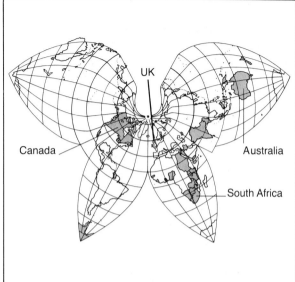

Figure 1   *The British Empire in 1920*

The benefits of Empire

The colonies did get some benefits from British rule. Modern developments began in farming, transport and education (Figure 2). But these changes were mainly to help the settlers stay in power and to exploit the country's resources.

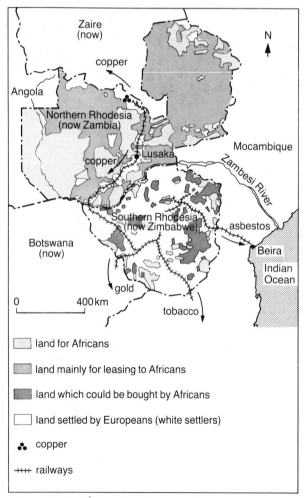

Figure 2   *Northern and Southern Rhodesia in 1953*

Exercises

(Heading: *Echoes of the past*)

1   a   What are colonies?
    b   Give three examples from Figure 1.

2   In what ways did Britain benefit from having an Empire?

3   Using Figure 2, copy and complete:
    Exports such as _____ and _____ were sent to Britain. Most of the land in Southern Rhodesia (now called _____) was taken over by _____. But there were _____ times more Africans in Southern Rhodesia. One benefit to the Africans was the building of _____.

The hamburger connection

In Central America, forests are being chopped down and land is being planted with grass (Figure 3). This is so that large herds of cattle can be fattened. The beef is mostly exported to the USA for use as luncheon meat, hot dogs and hamburgers.

A number of people benefit from this. The people of the USA get cheap beef, the local ranchers make money, and some local people get jobs. But the price of cutting down the forest is high (Figure 4).

Figure 3   *Rain forest in Central America*

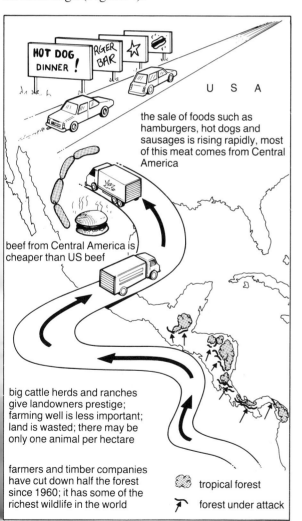

the sale of foods such as hamburgers, hot dogs and sausages is rising rapidly, most of this meat comes from Central America

beef from Central America is cheaper than US beef

big cattle herds and ranches give landowners prestige; farming well is less important; land is wasted; there may be only one animal per hectare

farmers and timber companies have cut down half the forest since 1960; it has some of the richest wildlife in the world

tropical forest

forest under attack

Figure 4   *The hamburger trail*

Exercises

(Heading: *The hamburger connection*)

4   Use Figure 4 to explain why forests are being destroyed in Central America.

5   Explain how each of the following statements shows why the forest should not be destroyed.
   a The rain forests of central America are rich in wildlife: a small country like Costa Rica has more species of birds than the whole of the USA and Canada.
   b One beef animal grazes 5 to 7 ha of cleared forest.
   c Local people are eating less meat each year (less meat than a cat does in the USA!).
   d Central America provides only 1·5% of US meat: Australia and New Zealand supply most beef imports and could supply more.

105

Dealing in death

The supply of weapons and drugs to poor countries is big business (Figure 5). Thousands of jobs in rich countries depend on this trade. Governments and transnationals spend millions of dollars making these products (Figure 6). The only way to make a profit is by selling as many as possible.

Figure 5 *Spending for peace*

Wasteful weapons

Super-powers like the USA and USSR sell weapons to governments who they support and want to keep in power. Some countries are ruled by people who do little to help the poor. The weapons are used to keep control over unrest.

Poor countries do need weapons to defend themselves, but the costs are staggering. Vast sums are spent which could be used to feed and help the poor (Figure 7).

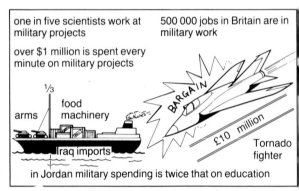

Figure 6 *The cost of weapons*

Exercises

(Heading: *Dealing in death*)

6   Copy out and match these statements with the facts in Figures 6 and 8
   the arms trade is big business
   Third World countries spend a large part of their precious money on arms
   arms are expensive
   many jobs in rich countries depend on the sale of arms to Third World countries
   many products sold are not essential to poor people
   transnationals do not keep the same high standards of quality as they do in rich countries

7   Study Figures 6 and 7 and then explain how arms supplied by rich countries are keeping people poor. Mention:
   costs...keeping governments in power.

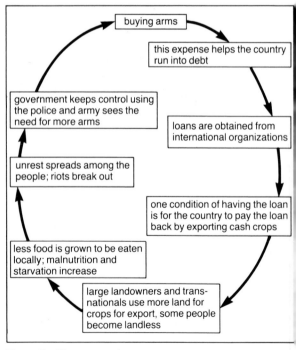

Figure 7 *The arms cycle*

Drugs which damage

Drugs used in medicines have brought great improvements in health. Most are supplied by transnational companies who can afford the research needed.

But some drugs are causing concern. They are expensive and not always suited to people's needs. Some countries such as Bangladesh are trying to control drug sales. They are making their own and making lists of suitable drugs. Many drugs sold by transnationals have been excluded (Figure 8).

Cigarettes and alcohol are new threats to health in poor countries. Drug companies use high-powered advertising to sell their products, sometimes in ways not allowed in rich countries.

Exercise

8 Copy out and match these statements with the facts in Figure 8.

a Many products sold are not essential to poor people.

b Transnationals do not have to keep the same health regulations as they do in rich countries.

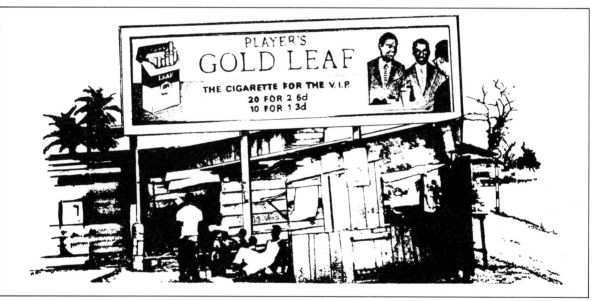

Figure 8 a *Cigarette advertising*

the sale of drugs and medicines in Bangladesh is worth £35 million

eight transnationals control 80% of the trade

a group of experts found that over 70% of drugs sold were harmful or not essential

there are no health warnings on cigarette packs

cigarettes are sold with 3 or 4 times more tar than in Britain

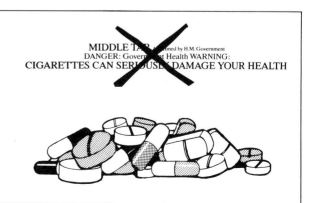

b *Facts about the drug and medicine trade*

5.5 Partners in trade

The world is an international market place. There is a battle for advantage between buyers and sellers. In this battle, the rich countries pick up bargains and become richer. People in poor countries are the losers.

Buying and selling

Poor countries supply raw materials (**primary products**) to rich countries. Without these raw materials, rich countries could not continue to manufacture goods. Some of these goods are exported to the Third World (Figure 1).

The pattern of trade

The poor countries do badly out of this **pattern of trade**. Prices of manufactured imports have gone up rapidly. But prices for primary products have not risen by much and often vary unexpectedly (Figure 2).

Unfortunately, many poor countries rely on a few products for their income (Figure 3). Poor harvests or a slump in world prices means an immediate cut in poor people's income.

Exercises

(Heading: *Patterns of trade*)

1 What are primary products? Give 3 examples from Figure 1.

2 Describe the pattern of trade shown in Figure 1.

3 What are the problems for poor countries shown in Figure 2?

4 Why are poor countries important to the rich world?

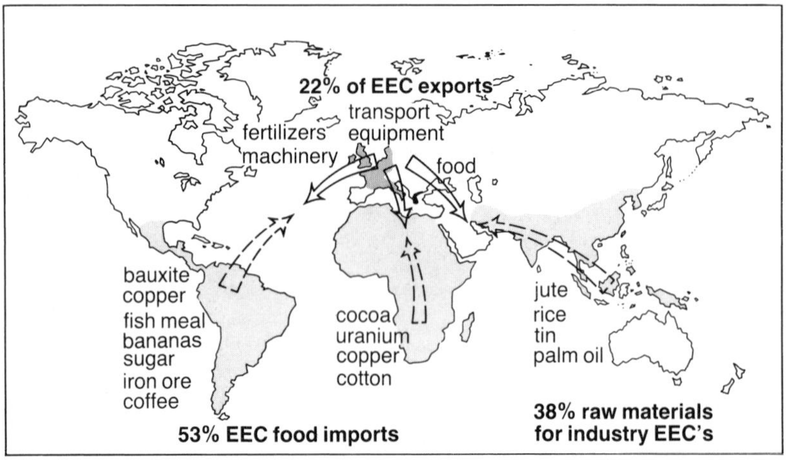

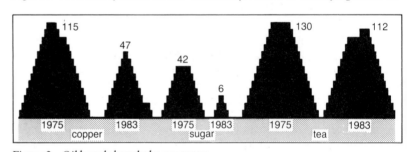

Figure 1   *EEC (European Economic Community) trade with developing countries*

Figure 2   *Oil barrels bought by one tonne*

country	main products	value of exports
Chile	copper (59), iron ore, fish meal	68%
Guyana	sugar (44), bauxite, rice	80%
Chad	cotton (60), beef, hides	68%
Zambia	copper (92), zinc, lead	98%
Burma	rice (50), timber, tin	

developing countries have important reserves of the world's tin, bauxite and nickel

developing countries supply the EEC with:
90–100% uranium ores
80–90% cocoa

Figure 3   *The importance of primary products*

Big business

The bulk of the Third World exports are bought by the governments of rich countries and by transnationals. They also do most of the transporting and selling. This buying and selling of products is called **marketing**.

Some transnationals also produce the exports. They have mines, ranches and estates in developing countries. Geest for example, have banana plantations in Central America. Unilever own palm oil estates in Nigeria. They pay taxes to the countries they operate in, but they also set wage rates and prices to suit their own interests.

Tightening control

The rich countries are taking more control over world trade. In farming, some transnational companies now control the whole system from crop to shop. They produce seeds, grow crops on land they own, supply farming equipment, then market the harvest (Figure 4). This system of organization is called **agribusiness**.

Exercises

(Heading: *Agribusiness in control*)

5 a What is marketing?
 b What is agribusiness?

6 Imagine you are a farmer who wants to:
  use seed which grows well without chemical fertilizers
  sell cassava to Europe
  use machinery which is made in his own country (*hint*: prices and spares)
 Why would you have problems?

7 Supermarkets sell a bigger variety of foods than ever before. How could food companies give us even more choice (Figure 4)?

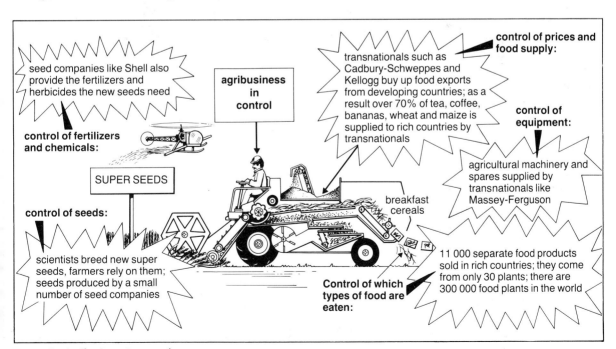

Figure 4 *Agribusiness in control*

Needing each other

One in six jobs in the USA depends on selling exports to developing countries. These countries also buy 33% of the Common Market's exports, and 45% of Japan's exports.

Both rich and poor nations depend on each other for both jobs and wealth. They are **interdependent**.

Competition from the poor

More and more developing countries are able to manufacture goods which they used to buy from rich countries. They are successful in the footware, electronics and textile industries. Such industries in the developed world have been badly hit by competition from the Third World (Figure 5).

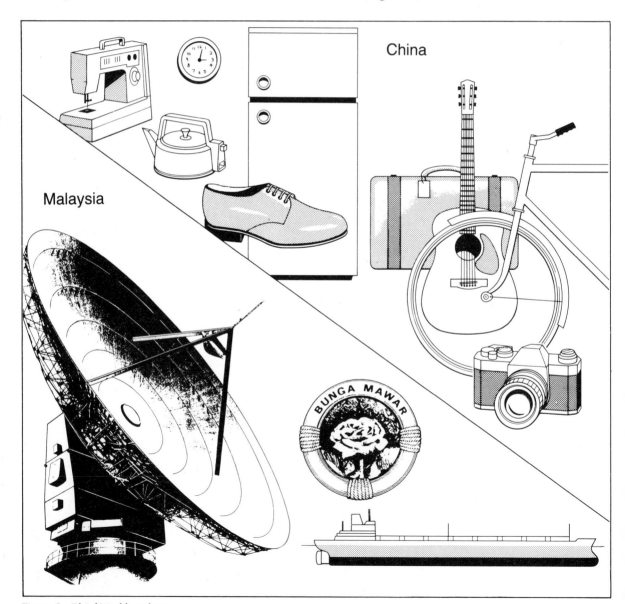

Figure 5 *Third World products*

Working for one world

The futures of the rich and poor peoples of the world are closely linked (Figure 6). But there is disagreement about how best we should trade and work with each other. The **Brandt report** said that the present system could work if some changes were made. Other people, especially in the Third World, want more drastic changes. For them, a **New International Economic Order (NIEO)** is needed (Figure 7).

Figure 6   *The world ship*

Exercises

(Heading: *Needing each other*)

7   a   Use Figure 5 to list goods made by developing countries. Add your own examples.
  b   Why is it important to rich countries that the poor become better off? (*Hint*: US jobs, Common market exports)

8   Study Figure 6.
  a   Look at each of the comments in turn. What is each saying about the real world?
  b   Try drawing your own version of the world ship as you think the world should be.

9   Study Figure 7.
  a   Sort out the statements into their correct pairs.
  b   Which two statements show that control is in the hands of the rich?
  c   What two major changes will Third World governments have to make if the NIEO is to be successful?

Today

1   the IMF (International Monetary Fund) and the World Bank give loans for development and to help countries out of debt; the north controls 60% of the votes

2   developing countries get heavily in debt and cannot pay back loans; Brazil owes $95 billion

3   huge sums are spent on paying for oil imports; Thailand uses up 40% of its money from exports in this way

4   rich countries are concerned that they will be flooded with cheap manufactures; they limit imports or else tax them to make them dearer

5   although more food is being grown in the world, some countries have less than a few years to go

6   trans-nationals control 30% of all that is grown, made and mined in the world; one third of the world's trade goes on between transnationals

Tomorrow

A   have a 12 million grain reserve in case of bad harvests and famines

B   lower trade barriers so the south can sell its manufactured goods more easily in rich countries

C   the south should have a greater share of the votes in organizations such as the World bank

D   control oil prices so they don't get too high or low; rely less on oil by looking for new energy sources such as hydro-electricity

E   rich countries to give a greater proportion of their wealth for aid; they have promised but not given 0·7%

F   more control of transnationals

Figure 7   *The world economic scene*

5.6 Fit to live in

The Arawak Indians who lived there called it Xamayca, a word meaning land of wood and water. Today it is called Jamaica, an island country in the Caribbean sea.

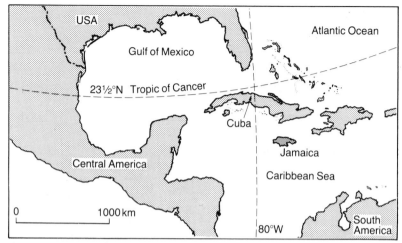

Figure 1 *Location of Jamaica*

Jamaican land use

Land use figures show how Jamaica has changed (Figure 2). Centuries of plantation farming has cleared the trees from most of the lowland. Peasant farming in the hills has cleared much of the rest. More recently, industry, mining and towns have continued to change the **environment**.

	%
woods	43·5
farming	46·5
natural grass	3·5
swamp	2·0
mining	0·5
urban	4·0

Figure 2 *Land use in Jamaica*

The pace of change

Changes are now happening faster than ever before. There are two main reasons for this. Firstly, the population is increasing by 1·2% each year. This means a need for more space and more resources such as water and wood. The population will double in about 30 years.

Secondly there is the scale and speed by which modern technology can change an area. Bulldozers and diggers can flatten slopes and dig out giant holes. Accidents such as an oil spill can cause widespread damage very quickly.

Exercises

(Heading: *Jamaican environment*)

1   Explain the word environment.

2   a   Graph the figures for land use in Jamaica (Figure 2).
    b   Explain why population and technology are changing the landscape faster than ever before.

3   Use Figure 3 to list Jamaica's natural resources. Use these headings:
       rocks and soil
       vegetation
       animals
       scenery
       climate

Figure 3 *Jamaica's natural resources*

Economy and resources

Jamaica is a small country of only 11 000 km². This is five times smaller than the UK. Development plans rely on using the Island's resources. The problem is how to do this without ruining the environment. There is no room to make mistakes on such a small island.

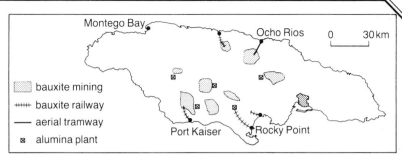

Figure 4 *Bauxite mining areas*

Bauxite mining

Selling bauxite earns 74% of Jamaica's export cash. Bauxite is the rock used to make aluminium.

About 16 200 ha is being mined for bauxite at the moment. Over the next 35 years, there are plans to mine up to 7% of the whole island (Figure 4).

Figure 5 *A bauxite quarry*

Bauxite and the environment

Bauxite is mined in open quarries (Figure 5). Next it is partly refined to make alumina, then exported.

The mining destroys trees and other vegetation. Rivers are sometimes disrupted when a new quarry is dug out. The biggest problem is caused by red sludge left after alumina is made. There is one tonne of waste for every tonne of bauxite mined. This gives 20 000 tonnes of waste every day.

The sludge is left to dry out in old pits and hollows. Some of the water sinks into the ground and mixes with rivers and underground water supplies. Polluted water affects farming, forestry, fishing, as well as drinking water.

At the ports, spills of rock cause problems to marine life such as fish, turtles and coral reefs.

Exercises

(Heading: *Resources and the environment*)

4 a What fact tells you that bauxite mining is important to the Jamaican economy?
   b List some of the uses of aluminium.
   c Why is it likely that bauxite will be mined for many years to come?

5 Use Figure 5 and text to describe a typical bauxite quarry in Jamaica.

6 Draw a diagram to show how the environment is affected by bauxite mining. Use sketches and labelled boxes to show how different parts of the mining operation affect rivers, vegetation, landscape and sea.

A complex web

Each type of development affects the environment in some way. But change in one part of the environment affects other parts. More mud in a river for example, affects wildlife, vegetation, and even the speed and course of a river. A system of links like this is called an **ecosystem**.

Around the coast

The coastal ecosystems of Jamaica are being affected by many types of development (Figure 6). Too many tourists can ruin the environment they have come to enjoy. Mining more bauxite means more money, but can make the Island less attractive to tourists. At the same time as looking after the environment, problems of high unemployment and poverty have to be solved.

Poverty the enemy

Poverty is the environment's main enemy. In rural areas, poor people cut down trees to make space for fields. This causes soil erosion and destroys wildlife. The Jamaican iguana for example, is now thought to be extinct.

In towns, shacks are built in areas with no water or sewerage facilities. This causes pollution to rivers and seas.

The best of plans go wrong when people's main aim is day to day survival.

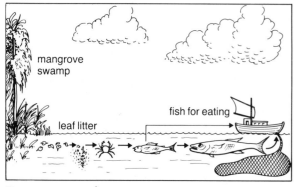

Figure 6   *A coastal ecosystem*

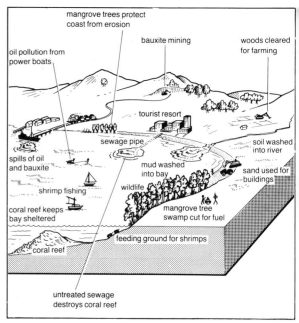

Figure 7   *The changing coast*

Exercises

(Heading: *Disturbing the ecosystem*)

7   Explain how each of the economic activities and natural features below can affect each other:

a tourism	coral reef
b mangrove swamp	erosion
c mining	fishing
d building	beaches
e fuel	wildlife

8   Describe the pressures on the environment caused by poverty in:
a   rural areas
b   urban areas

9   Solving problems of poverty in Jamaica may mean attracting even more tourists, mining for more bauxite and clearing more land for farming.
a   What problems can you see in these solutions?
b   Is there another solution you can think of?

One among many

Jamaica is one small country among twenty others in the Caribbean area. Twelve others share the surrounding mainland coast of North, Central and South America (Figure 8).

What happens in one country can affect the others. Rivers carry industrial waste, sewage and eroded soil and dump them in the sea. Currents spread the pollution from one country to another. The air is affected by forest burning and factory fumes.

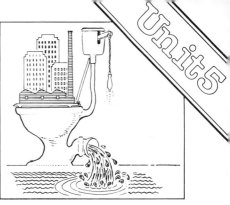

Figure 9 *Spreading pollution*

Different ideas

Different countries have different ideas on how best to become developed. Some want to explore for oil and build up industries. Others see tourism or agriculture as more important. These different ideas can conflict with each other. It is also true that while some governments care about the environment, others do not.

The wider world

Areas of tropical rain forest and mangrove wetlands are disappearing everywhere. Whole ecosystems are being destroyed and cannot be replaced. Humans are part of the global ecosystem. This is a good reason why care of the environment is important.

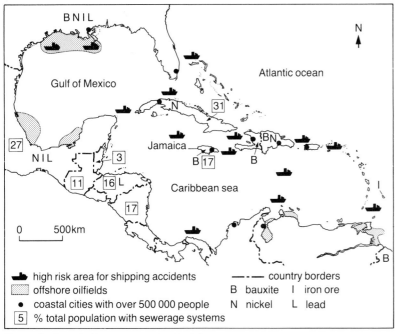

Figure 8 *Jamaica's position in the Caribbean*

Exercises

(Heading: *The need to co-operate*)

10 a Most of the countries in the Caribbean area are poor. Explain how this is likely to cause problems to the environment. Mention these things:

the cost of pollution control

the need for more farmland

the effectiveness of government laws

the need for rapid development

b Explain why pollution can also come from rich countries such as the USA. Mention:

the scale of technology being used

oil and fertilizers

factory pollution

11 a What are the different ways in which pollution can spread in the Caribbean? Give examples for your answers.

b Explain the importance of the statement: Humans are part of the global ecosystem.

5.7 Tourism, bonanza or blight

People on holiday spend money. In 1982, there were 290 million tourists spending money in another country. This is twice as many as there were ten years earlier. Some countries now rely on money from tourists to pay for their development plans. Tourism also has a darker side which brings problems.

Tourism and development

Of the 290 million **international tourists**, one in three visited a developing country. Like making cars or television sets, **tourism** is an industry which employs people and brings in money. This money can help pay for new factories, farm improvements, health and education services, or whatever else a government wants to spend money on.

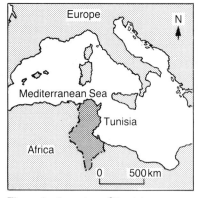

Figure 1   *Location of Tunisia*

Sun and sand

The beaches of Spain, Italy and Greece are well known tourist spots. Now countries in North Africa are trying to get a share in the business.

Tunisia has plenty of sand and sun (Figure 1). There are 1400 km of Mediterranean coastline. Inland there is the Sahara desert!

Money from tourism is very welcome. Average incomes in Tunisia are low. Few people work in industry and farming makes little money (Figure 2).

size	164 000 km²
population	6·7 million
GNP per head	$1310
income per head	US$1246
% in farming	32

Figure 2   *Tunisian statistics*

temperature in °C	J	F	M	A	M	J	J	A	S	O	N	D
sunshine in hours	9·2	10·8	14·4	16·7	19·8	23·3	25.3	25·6	24·7	21·7	14·7	13·6
rainfall in mm	4·5	5·9	6·4	7·7	9·5	10·5	11·1	10·7	9·6	8·5	6·2	4·5

Figure 3   *Tunisian climate*

Figure 4   *A hotel by a Tunisian beach*

Exercises

(Heading: *Tourism in Tunisia*)

1 a Make a copy of Figure 1.
   b On your map, mark how far the UK is from Tunisia. Use an atlas.

2 Use Figures 2, 3 and 4 to write an account of the attractions of Tunisia to tourists from countries such as the UK. Use the headings:
   location   climate   coast   inland

3 What facts about Tunisia show that it is a poor country where money from tourism would be useful?

Tourists to Tunisia

Figure 5 shows how the tourist industry has grown and where tourists come from. More tourists mean more hotels and more jobs. Money from tourists is now the second most important industry earning money from other countries (**foreign exchange**).

Tourism Tunisian style

Tunisia tries to offer something different. Tunisia has its own Arab culture with foods, dance, handicrafts and festivals. Hotels are being built to fit in with local building styles to keep the atmosphere Tunisian (Figure 6).

Sea and Sahara

The Ministry of Tourism offers: sea and Sahara in one day. The plan is to spread tourist centres around the country, both on the coast and inland. This is to bring the benefits of tourism to as many areas as possible. The tourists will also benefit by being able to see more of the country (Figure 7).

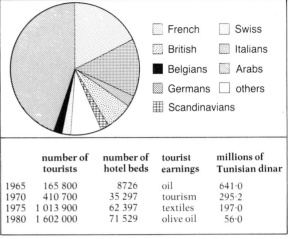

French · Swiss · British · Italians · Belgians · Arabs · Germans · others · Scandinavians

	number of tourists	number of hotel beds	tourist earnings	millions of Tunisian dinar
1965	165 800	8726	oil	641·0
1970	410 700	35 297	tourism	295·2
1975	1 013 900	62 397	textiles	197·0
1980	1 602 000	71 529	olive oil	56·0

Figure 5   *Visitors to Tunisia*

Figure 7   *A day trip to the Sahara*

Exercises

(Heading: *Planning the tourist industry*)

4   Use Figure 5 to describe the way in which tourism has grown up in Tunisia. Write a paragraph about the information. Be sure to mention these things:
      the date of the figures
      the overall trend of the figures
      has change been slow or quick?
      where tourists come from
      the importance of tourism to the country

5   Draw a page of diagrams and information to show how a holiday in Tunisia tries to offer something different.

Figure 6   *A Tunisian hotel*

Money that leaks

Figure 8 shows that tourism can both cost as well as earn money. Most international tourists are from rich countries and expect the highest standards when on holiday. This means the best hotels with modern facilities.

These hotels are usually owned by transnational companies based in the UK, France or the USA. The holiday will probably be organized by a tour company or airline also based in the tourist's home country (Figure 9).

Most of the cost of the holiday is in the air fare and hotel. Profits from these go back to the rich country. The poor country is left with between 30% and 40% of the total.

Waiters and cleaners

The highest paid jobs go to executives from the transnational's home country. Local people are employed in the low-paid, unskilled jobs such as waiters and cleaners.

Because tourism is **seasonal**, many workers are laid off for several months each year.

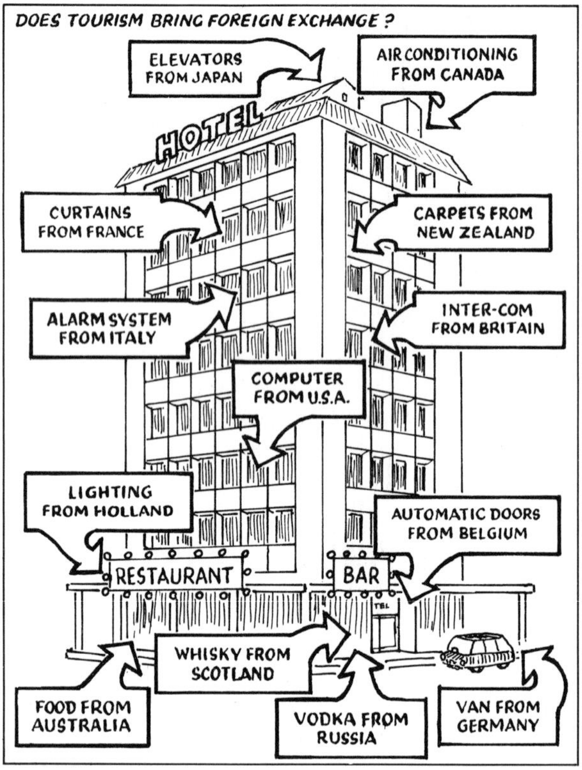

Figure 8   *Money that leaks*

airlines share in world market top 10 airlines	% share
British Airways	6·5
Pan American	6·5
Japan Air Lines	3·8
Air France	3·7
Trans World Airlines	3·3
Lufthansa	3·2
KLM Royal Dutch Airlines	2·9
Quantas	2·7
Alitalia	2·4
Swissair	2·1

hotels owned by transnational companies country	number of hotels	number in developing countries
USA	223	112
France	39	23
UK	10	1
Japan	30	6

Figure 9   *Who benefits from tourism?*

Exercises

(Heading: *Problems from tourism*)

6   a   Use Figures 8 and 9 to explain why building up a tourist industry can be expensive to a poor country. Why is the same not true for a developed country?

   b   Explain why so much money from tourism does not stay in the developing country. Mention these things:
   who organizes the holiday
   how a holiday is paid for
   how a tourist spends money on a holiday

7   Tourism does provide jobs, but there are two problems in the type of work involved. Explain what these problems are.

A clash of cultures

Growth in tourism has also brought social problems. The poem and diagrams (Figures 10, 11 and 12) come from Thailand in SE Asia. Thailand is visited by rich tourists from Japan, the USA and Australia. They come to a country where the culture and traditions are very different and the people are very much poorer.

Tourists stay in luxury hotels while local people live in poverty. Local customs and shrines lose their real meaning as they become stopping points on a tourist's day out.

Problems of prostitution have reached serious levels. Poverty drives some people to make money in whatever way they can.

Alternative Tours Ltd. (A Thai tour company)

Perhaps an answer is not to allow so many tourists to come. One small tour company also tries to make people understand what they see. They try to show visitors the country as it really is, away from the tourist centres. They hope this will make the holiday more enjoyable for both the visitor and those being visited.

> And if the soul of this land
> is behind the tourist poster
> beckoning to sun, sea & sand,
> it is equally there in the gutter
>
> where beggars fight off stray
> cats for the slop of left-over
> dinners, where mice foray
> offal cast by itinerant hawkers.
>
> It is there on the peeling
> alley walls weeping nicotine-
> flecked globs of phlegm, reeking
> putrescent fruit, faeces & urine.
>
> Cheap pineapple and tropical
> splendour you now enjoy, dear
> traveller, is paid with impossible
> lives lived out in unspeakable squalor.
>
> C. Rajendra

Figure 10   *A poem about tourism*

Figure 11   *An advert for escorts*

Figure 12   *Contrasts in wealth*

Exercises

8   Read the poem in Figure 10. What is the poem saying to rich tourists about the cost of their holiday?

9   What do Figures 11 and 12 tell you about the social problems caused by the tourist industry?

10 a What are the disadvantages in allowing fewer tourists to come to a developing country?
    b What do you think tourists might think about the idea put forward by Alternative Tours Ltd?

5.8　The final frontiers

Poor countries are in a race to become more developed. They are trying to do quickly what the richer countries took two centuries to do.

Opening up new land

Poor countries need to earn money as quickly as possible. One way to do this is to find more resources such as minerals. These can either be sold to rich countries, or used to build up new industries.

This means **opening up** areas which so far have been untouched. These areas are usually remote and difficult to live and work in, such as islands, deserts, rain forests and mountains. Rich countries can supply the technology to build roads, airfields and mines, even in places that were once hard to get to (Figure 1).

Growing population numbers means there is a need to grow more food. Finding new areas to farm is one answer to this problem.

The right decisions?

Using up resources and opening up new land seem to be quick and easy answers to development problems. But untouched places become fewer every year (Figure 2).

Many resources, once used, cannot be replaced. The rich countries have already shown how easy it is to waste resources and destroy the environment for the future.

Exercises

(Heading: *Opening up the final frontiers*)

1. a What is meant by the term opening up?
   b What kind of areas make up the last frontiers?
   c Why have some areas remained unchanged for so long?
   d Explain why the few remaining unchanged areas are no longer safe from development.

2. Write a paragraph using the information in Figure 1 to explain what the pressures are to open up new areas.

3. Developed countries have not always used their resources wisely. What lessons can developing countries learn about using resources and caring for the environment?

Figure 2　*Vanishing wildlife*

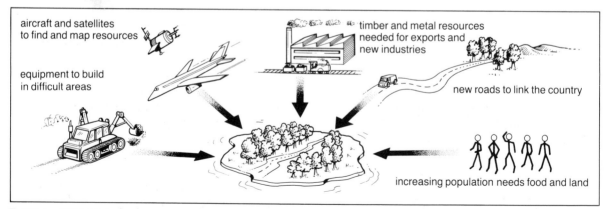

aircraft and satellites to find and map resources

equipment to build in difficult areas

timber and metal resources needed for exports and new industries

new roads to link the country

increasing population needs food and land

Figure 1　*Pressures to develop*

Native peoples

There are some places where very few people live. There are only 100 000 Indians in Brazil's vast Amazon forest and 50 000 bushmen in southern Africa. Groups such as these are called **indigenous** people. They have lived in these areas for hundreds of years.

People and their environment

About 18 000 indigenous people live on the Indonesian island of Siberut (Figure 3). Most live in small groups. They hunt, collect, fish, grow some crops and rear pigs (Figure 4).

Compared with the technology of a developed country, this way of life is said to be simple or **primitive**. But it allows people to survive in difficult conditions, without destroying their environment. Their way of life depends on careful use of the environment.

In the way

Very few people live in isolation any more. Missionaries, health workers, traders and government advisers all try to change the old ways (Figure 5). They bring new tools, clothes, beliefs and often diseases. Native people sometimes get in the way of mining and schemes for new farm land. With no legal papers, they are moved on and do not share the wealth. Decisions are made in a far-off capital city. The people themselves have no say in what happens to them.

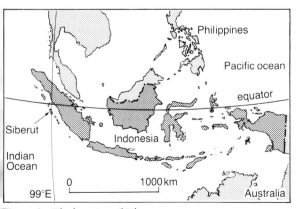
Figure 3   *The location of Siberut*

Figure 4   *A primitive way of life*

Figure 5   *The native problem*

Exercises

(Heading: *Indigenous peoples*)

4   a   Explain the term indigenous or native people. Name some examples.

   b   Describe the traditional way of life of these people, for example, the Siberut islanders. Explain why this way of life is called simple.

   c   Why is it so important to indigenous people not to change the environment?

5   Imagine oil was found in a remote area where a small native group live.
   What do you think should be done?
      allow drilling to go ahead
      move the native people out of the area
      give the native people a share in the profits
      try to give the people jobs on the oil rigs
      any other ideas you have
   Why do you think your decisions are right?

5.9 Planning development

The development of Santa Cruz

Santa Cruz is an imaginary developing country. Imagine that you could plan its development over the next 30 years.

1 Make an accurate trace of the map of Santa Cruz on page 124.
2 Work *down* the page opposite starting with Period 1. Fro each box in turn decide what you want to do.
3 After deciding, use the key on the map to plot your developments on your map.
4 Continue in the same way with Periods 2 and 3. But if boxes are joined with an *arrow* then they both have to be plotted.
5 Now turn to the development scores. Total up your score.

Score over 300	– a promising future for Santa Cruz
120 – 300	– some decisions will need changing
below 120	– don't be surprised if there is a revolution!

6 Look at the development scores again. Are there any you would change? Why?

Development scores

15 revolutionary literacy campaign
5 government trained literacy campaign
1 each primary school
1 each clinic
5 hospital
5 each health worker

1 motoway
5 bridge
10 road or surfaced track
5 airport extended
15 green revolution
20 bank loans to poor farmers
5 limited land reform
25 widespread land reform

10 coal exports
10 coal mined
20 dam and electricity
20 electricity supplied to towns or villages
5 steel works
15 village crafts
10 tourist hotels
5 rest houses

activity	the present scene	period 1	period 2	period 3
education	few rural schools; more wealthy send their children to private schools in the city; 60% literate	little change (add nothing) or 5 villages build primary schools	little change (add nothing) or build 5 more primary schools in villages	government trains people to lead literacy campaign or revolutionary government uses students and children to teach adults to read and write
health	widespread malnutrition; outbreaks of cholera	5 scattered village clinics run by missionaries	little change (add nothing)	large hospital built in capital or 30 health workers trained
transport networks	mostly tracks and footpaths; only surfaced roads link main towns	100 kilometres of new motorway built or rural roads built to link 5 villages	airport extended to take tourist flights ⇨ or 100 kilometres of track given a hard surface ⇨	150 kilometres of new motorways built or rural roads built to link 25 villages
farming	80% people work on land; have small subsistence plots; most land owned by wealthy landowners	little change (add nothing)	green revolution throughout the country ⇨ or easy for poor farmers to get bank loans ⇨	some land reform (shade in area A) or widespread land reform (shade in areas A and B)
resources	two large rivers; large deposits of coal in the west	government asks foreign transnational to build hydro-electric dam ⇨ or open up new coal mine ⇨	electricity lines built to main towns ⇨ or rail link built to capital from coal mine and coal exported ⇨	20 villages get electricity or money from coal exports used to import cars and wheat for bread
industry	a few factories in main towns; process food and manufacture textiles	little change (add nothing)	iron and steel works built at capital ⇨ or help given to village crafts in 20 villages ⇨	no change or 10 more villages given help with crafts
tourism	few tourists; is an isolated country; no tourist amenities	little change (add nothing)	3 hotels for tourists in towns ⇨ or 3 rest houses for tourists in national park ⇨	6 villages make crafts for tourist trade or 20 kilometres of road built in national park

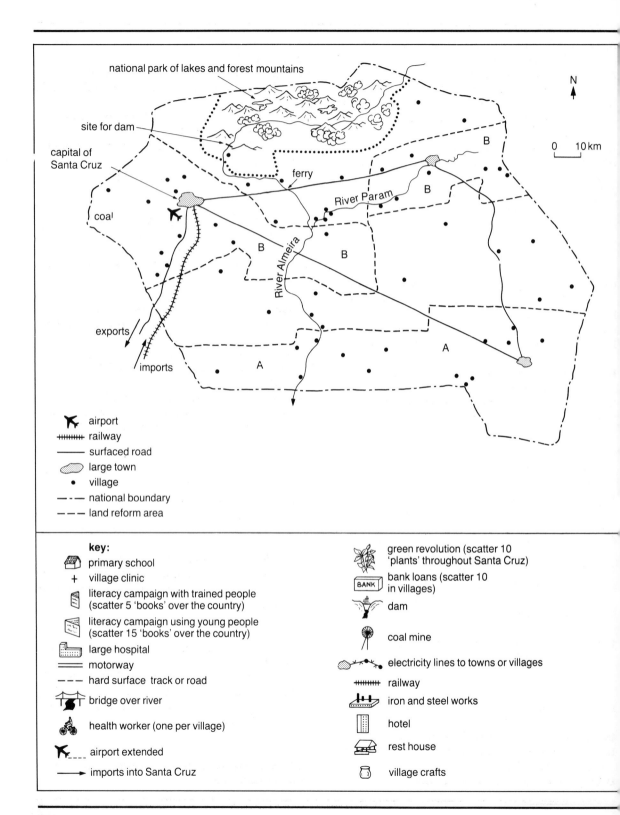

national park of lakes and forest mountains

site for dam

capital of
Santa Cruz

coal

ferry

River Param

B

B

B

B

River Almeira

A

A

N

0   10 km

exports

imports

key:

✈ airport
╫╫╫ railway
── surfaced road
☁ large town
• village
—·— national boundary
——— land reform area

🏠 primary school
+ village clinic
📖 literacy campaign with trained people
   (scatter 5 'books' over the country)
📖 literacy campaign using young people
   (scatter 15 'books' over the country)
🏢 large hospital
═══ motorway
––– hard surface track or road
🌉 bridge over river
🚴 health worker (one per village)
✈··· airport extended
──▶ imports into Santa Cruz

🌿 green revolution (scatter 10
   'plants' throughout Santa Cruz)
🏦 BANK bank loans (scatter 10
   in villages)
⚙ dam
⚙ coal mine
☁••• electricity lines to towns or villages
╫╫╫ railway
🏭 iron and steel works
🏨 hotel
🏠 rest house
🏺 village crafts

Glossary

Agrarian reform: Changes to improve farming and the way the land is owned/2.6

Agribusinesses: Large companies involved in farming and supplying its needs/5.5

Agricultural extension workers: People trained to go into rural areas and give advice on farming methods/2.5

Aid: Help given by one country to another/5.2

Aliens: People living in a country who do not have the nationality of that country/4.4

Alluvium: Mud and silt left in a valley bottom as a result of flooding/2.2

Appropriate technology: Machinery, often low cost, suited to local conditions/3.2

Barefoot (doctors): People trained to work in local communities using simple medical techniques/4.2

Bilateral aid: Aid from one country's government to another/5.2

Birth control: Methods used to prevent children being born/4.2

Birth rate: The number of children born in a year for every 1000 people/1.4

British Empire: The name given to places when they were ruled by Britain/5.4

Capital intensive: Using expensive machinery rather than people/3.5

Cash crops: Crops grown to sell for money/3.4

Colonies: Countries which were once ruled by another country/5.4

Communist: A system of government in which the state has control over most aspects of life/4.2

Co-operative: People working together/2.6

Core regions: A more developed part of the country, often containing the capital city/4.6

Cottage industries: Small businesses carried on in people's homes/3.1

Cut and burn: A farming system where forest is cut down, burnt, then farmed for a year/4.3

Death rate: The number of deaths for every 1000 people in a year/1.4

Deforestation: The large scale removal of forests/4.3

Democratic: A system of government where people vote for their rulers/4.2

Demography: The study of population/1.4

Densely populated: A large number of people living in an area/4.3

Dependancy load: The number of young children and old people in a country who rely on the people at work/1.4

Desertification: Making a desert, usually through man's overuse of land/4.3

Developed: Having modern farming, industry and services/1.2

Development: The process of building up modern farming, industry and services/1.2

Developing (countries): Poor countries which are building up their farming and industries and are improving the standard of life for their people/1.2

Drought: A long period without rain/1.5

Dryland farming: Methods of farming in areas with little rain/2.2

Economy: The way a country earns money/1.2

Ecosystem: A system of links in the environment/5.6

Elite: A group of privileged people/4.6

Energy: Types of power/1.2

Environment: The features of an area/5.6

Eradicated: Removed for ever/2.3

Exports: Goods sent out of a country/1.2

Extended family: A family where relatives such as aunts, uncles and cousins live together/1.4

Fallowing: Leaving land uncultivated so that it can become fertile again/2.2

Flood-retreat farming: Farming land after flood levels have gone down/2.2

Foreign exchange: Money earned from other countries/5.7

Free trade zone: An area where normal tax laws are removed so goods can be made then exported/4.5

Freedom: Having the ability to live your life without undue interference/4.2

Green revolution: Scientific methods to grow more farm produce/3.4

Gross National Product: The value of all goods and services produced by a country/1.2

Gulley: Where soil has been removed on a hillslope in a deep V-shape/4.3

Health workers: Local people trained to carry out simple health care/2.3

High technology: Using advanced techniques in industry/3.3

High yielding varieties: Crops which produce a large amount of food/3.4

Imports: Good brought into a country/1.2

Income per head: The amount of money each person in a country earns on average/1.2

Index: A measure/1.2

Indigenous: People who have lived in the same place as their ancestors for a long time/5.8

Industrialized: A country which has built up its industry/1.6

Infant mortality rate: The number of deaths among very young children for every 1000 born in a year/1.4

Informal (sector): Jobs which people do from time to time, usually self-employed service jobs/4.4

Integrated rural development: A scheme involving improvements to both farming and living conditions/2.6

Intercropping: Growing several crops close together/3.4

Interdependent: When countries need each other/5.5

International tourists: Tourists who go from one country to another/5.7

Investing: Putting money into a project with the aim of making a profit later/3.5

Irrigated: Farmland that has been watered/3.5

Labour intensive: Using people to do work rather than machines/3.5

Landless: Not having any farmland to own or rent/4.6

Land-locked: Having no access to the sea/1.5

Land reform: Changes in land ownership/2.6

Least developed countries (LDCs): The poorest countries in the world/1.5

Life expectancy: How long people can expect to live/1.4

Low cost housing: Cheap simply built houses/4.5

Man-made (disasters): A disaster caused by something people do/2.3

Marketing: Advertising and selling goods/5.5

Mass vaccination: Giving a very large number of people a drug to prevent disease/2.3

Migration: Movement of people from one place to another/4.1

Models: A generalized way to describe how something happens/4.1

Monoculture: Growing only one crop/3.4

Monsoon (countries): Countries in Asia where there is a monsoon climate/1.5

Multilateral aid: Aid to a country which comes through an international organization or from a group of countries/5.2

Multinationals: Firms which have business interests in many countries/5.3

Multi-purpose scheme: A scheme which does many things at the same time/2.3

Natural (disasters): Disasters caused by an act of nature/2.3

Natural increase: The growth in population as a result of more births than deaths/1.4

Naturalization: To take the nationality of a country someone has moved to live in/4.4

Negative correlation: As more of one thing happens, less of something else happens/1.2

Newly industrialized countries: Countries which have built up industry very quickly/3.3

New International Economic Order (NIEO): A new arrangement for trading between countries/5.5

New town: A newly built town with houses, industry and services/4.5

North (and South): The world's rich countries/1.2

Opening up: To bring development to an area that has remained mainly unchanged/5.8

Overused: To use land so much that it loses its fertility/4.3

Pattern of trade: The balance of imports and exports between countries/5.5

Periodic markets: A market held from time to time at a convenient place/2.4

Plantations: A very large organized farm, usually growing only one product/3.4

Population control: Limiting the number of people in a country/4.2

Population pyramid: A drawing which shows the age and sex of a population/1.4

Population structure: The proportion of people of different ages in a population/1.4

Positive correlation: As one thing increases, something else also increases/1.2

Primary (employment): Jobs involving farming, fishing, forestry and mining/1.2

Primary health care: Looking after basic health needs/2.3

Primary products: Raw materials and foods/5.5

Primate city: A city many times larger than the next largest city in a country/4.5

Primitive: A way of life that has changed little over a very long time/5.8

Protein−energy malnutrition (PEM): Very poor health caused by lack of the right balance of foods/1.3

Quota: An amount that is allowed/4.2

Rain-fed (farming): Farming which relies on rain as the source of water/2.2

Raw materials: Things needed to make something from/5.5

Refugees: People who are forced to flee from one place to another/2.3

Resources: Things which can be used/3.6

Sahel: Countries along the southern edge of the Sahara desert/1.5

Savanna: Grass and scattered tree vegetation which grows near the tropics/2.2

Seasonal: Something which happens only at certain times of the year/5.7

Secondary (employment): Making things in a factory/1.2

Sectors: A group/1.2

Self-help: Things which people do themselves to improve their lives/2.5

Self-reliant: Dependent on doing things yourself/3.3

Semi-arid: Almost as dry as a desert/2.2

Shanty towns: Areas of poor housing usually on the outskirts of Third World cities/4.5

Share croppers: Farmers who pay rent by giving part of the crop to the landowner/4.6

Site and services (scheme): Building a basic plot for houses with water and sewerage/4.5

Slum upgrading: Improving houses which people have usually built for themselves, often in poor countries/4.5

Soil erosion: The removal of soil/4.3

South (and North): the world's poor countries/1.2

Squatters: People who live on land illegally/4.6

Standard of living: How well people are fed, their health and living conditions/1.2

Starvation: Suffering from a lack of food/1.3

State farms: Farms owned by a government/2.6

Subsistence: providing most of your own food/5.5

Super-powers: The powerful nations, USA and USSR/5.4

Tenants: People who rent land/4.6

Tertiary (employment): Jobs in the services such as shop and office work/1.2

Third World: The poor countries of the world/1.2

Tied (aid): Aid money which must be spent on goods or services from a particular country/5.2

Trade: The movement of goods from country to country/1.6

Traditional birth attendants: Women in rural areas who help at births/2.3

Transnationals: Firms which have businesses in many parts of the world/4.6

Trickle-down: The idea that the benefits of development will eventually help the poorest people in a country/5.2

Tropical rain forest: An area of forest growing between the two Tropics/4.3

Urbanization: The increasing percentage of people living in towns and cities/4.5

Underemployed: Only being able to find work from time to time/4.4

Undernourishment: A lack of the right kinds of food/1.3

Yields: The amount of food from a crop/3.4

Zero growth: No increase in a population/4.2